PLC Programming for Industrial Automation

Kevin Collins

British Library Cataloguing In Publication Data

A Record of this Publication is available
from the British Library

ISBN 978-1-84685-598-6

First Published 2007 by

Exposure Publishing, an imprint of Diggory Press,
Three Rivers, Minions, Liskeard, Cornwall, PL14 5LE, UK
WWW.DIGGORYPRESS.COM

Contents

Introduction

I have been teaching PLC programming for fifteen years and the question that I hear most often from students is "Can you recommend a book on this?" In response I have trotted out the titles of various standard text books but I have never come across a book that really develops the skill of PLC *programming* instead of telling the reader what PLCs are all about. I have finally decided to fill this gap in the market myself.

"What sort of PLCs do you use?" is another popular question. It implies that familiarity with one make and model of PLC will leave the programmer struggling when asked to use a different type. I deliberately teach a generic style of programming that allows the learner to switch between types of PLC as easily as between different makes of electronic calculator. Every skill needs practice however and my thanks are due to *TriLogi* for permission to use their excellent PLC simulator software throughout this book. The students can load the software onto a computer and practice the examples and exercises provided.

The third problem that authors have failed to address is the variety of programming languages available. Ladder logic is by far the most popular programming language in use because of its resemblance to hard-wire control diagrams. On its own, however it is unsuitable for complex programs. As the automation task grows so the ladder program expands organically, until only the original programmer can find his way through the tangle of inputs and outputs, relays and function blocks.

This problem has been solved by the use of Sequential Function Chart (GRAFCET) methods but the obvious popularity of ladder logic persists. The solution is to plan the program using a sequential function chart and then to enter it into the PLC using ladder logic. In this way program is highly structured, standardised and easy to debug and modify, while the familiarity of ladder logic is preserved.

The first two chapters of the book are used for programming basics. The remainder concentrates on the control of automation sequences commonly found in industry.

The examples used in the book have all been thoroughly tested and their suitability for use in the classroom and in industry established.

Chapter 1
PLC Basics

1.1 Function of a PLC

A PLC is a microprocessor-based controller with multiple inputs and outputs. It uses a programmable memory to store instructions and carry out functions to control machines and processes.

The PLC performs the logic functions of relays, timers, counters and sequencers. It has the following advantages:

Low cost
Reliability
Reprogramability

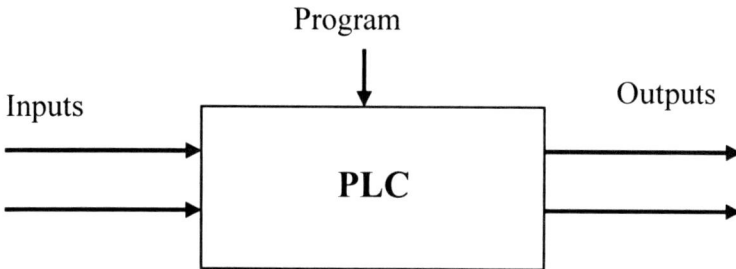

Program

Inputs   Outputs

PLC

Fig 1.1 A programmable logic controller

1.2 Inputs and Outputs

The PLC inputs give it information about the machine or process that it is controlling. These are typically switches and sensors. The switches are connected to an input module that provides the interface between the switches or sensors and the PLC.

Input module circuits have opto-isolators to protect the internal PLC circuitry from damage.

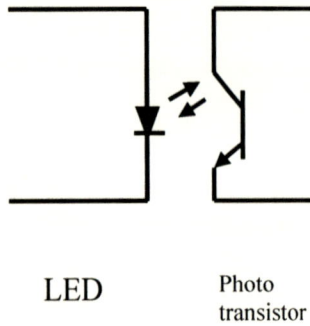

LED   Photo
transistor

Fig 1.2 An Opto-Isolator

The PLC outputs are connected directly or indirectly (e.g. through a relay) to actuator controls. Examples include solenoids on directional control valves, motors, motor contactors, alarms and warning lights.

There are three main types of output module:

Relay (volt-free): The signal from the PLC operates a relay within the output module connecting the control voltage to the output port and hence to the actuator.

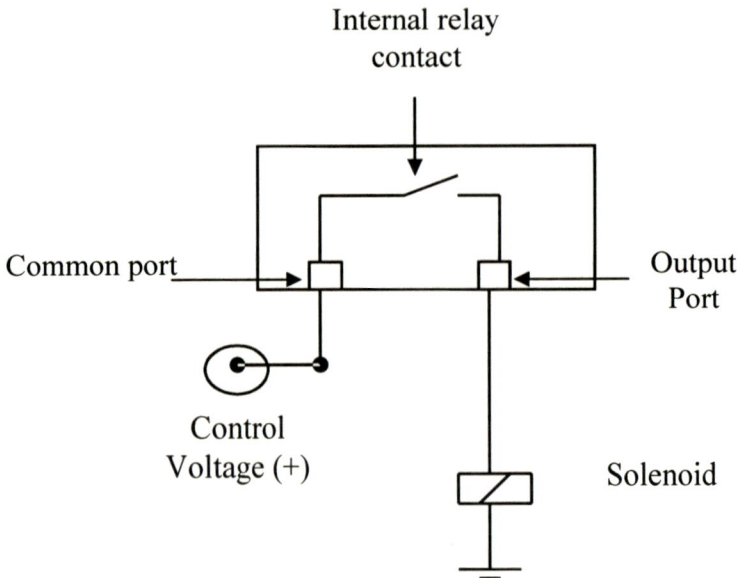

Fig 1.3 PLC Relay Output

Transistor: A transistor is used to switch on the output. This is faster than a relay output but is only suitable for low power direct current applications.

Triac: This solid state device is used for switching alternating current devices. It requires some form of over current protection.

1.3 PLC Architecture and Wiring Diagrams

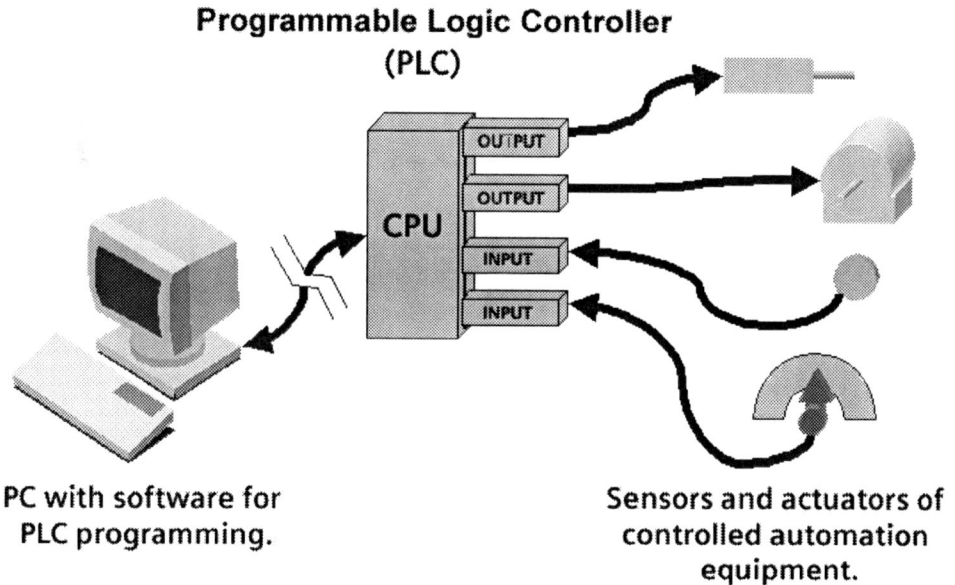

Fig 1.4 PLC Connections

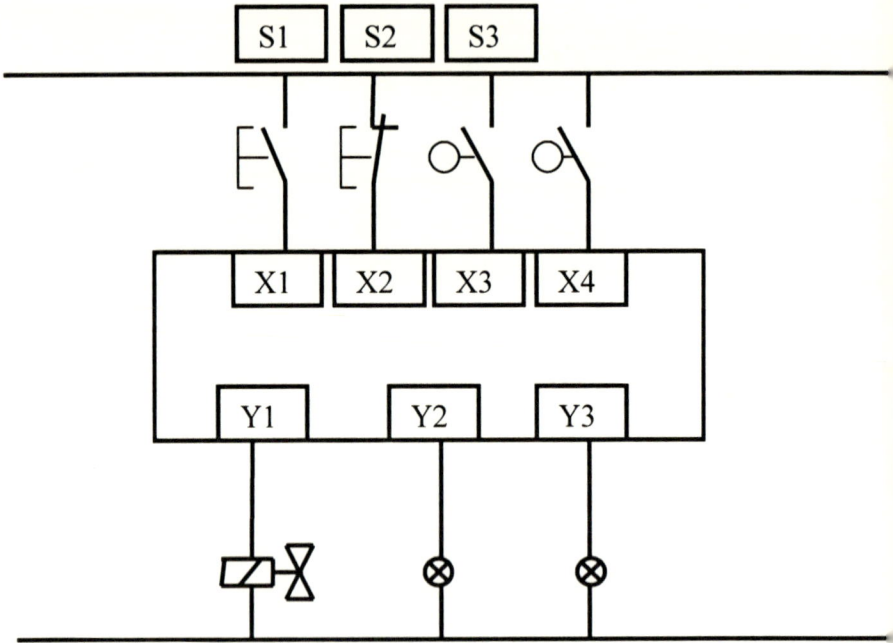

Fig 1.5 PLC wiring diagram

Fig 1.4 shows a pictorial view of the PLC with its connections. In practice we work with a simplified diagram as shown in Fig 1.5.

1.4 Network Protocols

The wiring diagram in *Fig 1.5* shows the inputs and outputs connected directly (*hard wired*) to the PLC. The devices shown are on/off or digital in nature but the signal to the PLC is analog. Many commonly used devices conform to a 4-20 mA standard whereby signals of 4mA and 20mA form respectively the minimum and maximum values of an analog signal.

With analog devices, a separate cable needs to be run between the end device and the control system because only a single analog signal can be represented on the circuit. The 4-20 mA standard is slowly being replaced by network or *fieldbus* communications. *Fieldbus* is a multi-drop digital two-way communication link between intelligent devices. *Fieldbus* allows the connection of a number of sensors all located in the same area to the same cable. *Fieldbus* comes in many varieties depending on the manufacturer and application. Examples include *ASibus*, *Profibus*, *Devicenet* and *Modbus*.

A more recent trend is the development of *Industrial Ethernet* which has the capacity to transport large quantities of data not only for process control but also to integrate the process with management information systems.

This book concentrates on PLC programming and while the sample wiring diagrams are of the type shown in *Fig 1.5* the programs are designed to receive data from inputs and to send data to outputs regardless of the network system being used.

Questions

1. Switches, proximity devices and sensors are generally used in what way in a plc application?

Answer:	⌒	a. Relays
	⌒	b. Software elements
	⌒	c. Inputs
	⌒	d. Outputs

2.

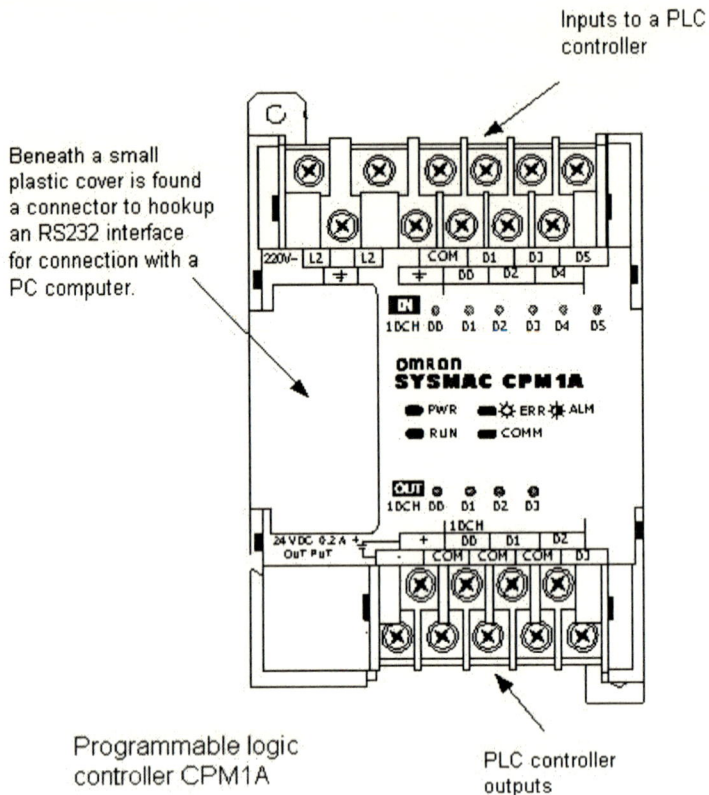

Inputs to a PLC controller

Beneath a small plastic cover is found a connector to hookup an RS232 interface for connection with a PC computer.

Programmable logic controller CPM1A

PLC controller outputs

Fig 1.6

In the diagram *Fig 1.6* of a plc. Why would it be necessary to connect a PC?

Answer:

    ◌   a. To read the inputs and set the outputs.

    ◌   b. To store the output values

    ◌   c. To edit the plc program.

    ◌   d. To store the input values.

3. Which option below best describes the action of an opto-coupler?

Answer:

    ○   a. It breaks the contact when there is excess current.

    ○   b. It breaks the contact when there is excess voltage.

    ○   c. It transmits the input signal using fibre optics.

    ○   d. It isolates the plc from the input voltage

4.

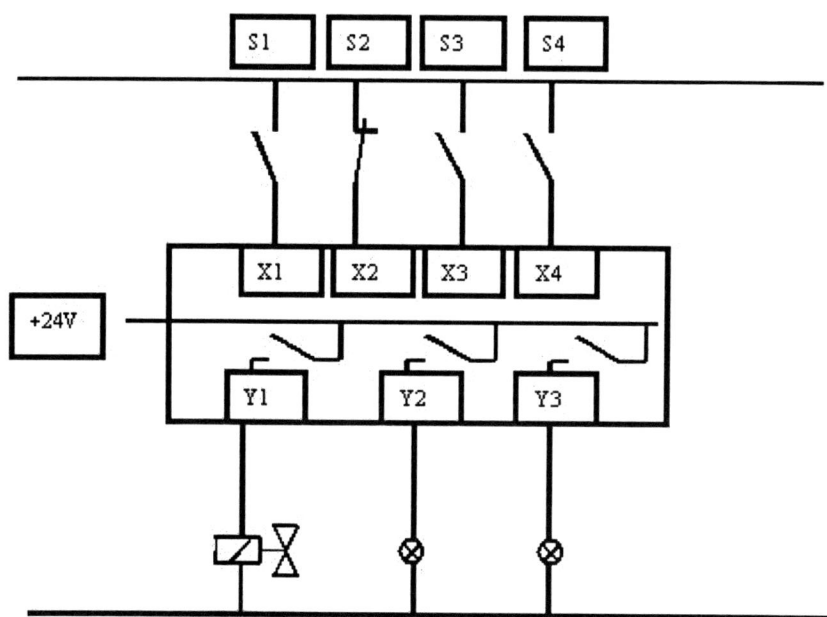

Fig 1.7

Study the diagram *Fig 1.7* and pick the correct statement about it.

Answer:

     ◌   a. When the power is switched on to the plc all the n/o relay contacts shown close.

     ◌   b. The 24 V supply shown is used to power the plc itself.

     ◌   c. The plc energises an output by closing the relevant relay contact.

     ◌   d. When the plc outputs are energised they are all latched on by the relay contacts shown

5. "This type of plc output is solid-state and is used for switching alternating current."

The description above best describes what type of switch?

Answer: ◌   a. triac

     ◌   b. push button

     ◌   c. transistor

     ◌   d. relay

Chapter 2
Ladder Programming

2.1 Conditional Logic

The PLC scans its inputs and, depending on the program, switches on or off various combinations of outputs. The logic state of the output depends on the input conditions and so the term *conditional logic* is used.

A simple example of conditional logic could be stated as follows:

"A machine switches on if either of two *start* switches are closed and all of three *stop* switches are closed."

The conditions could be realised by a hard wire solution as shown in Fig 2.1.

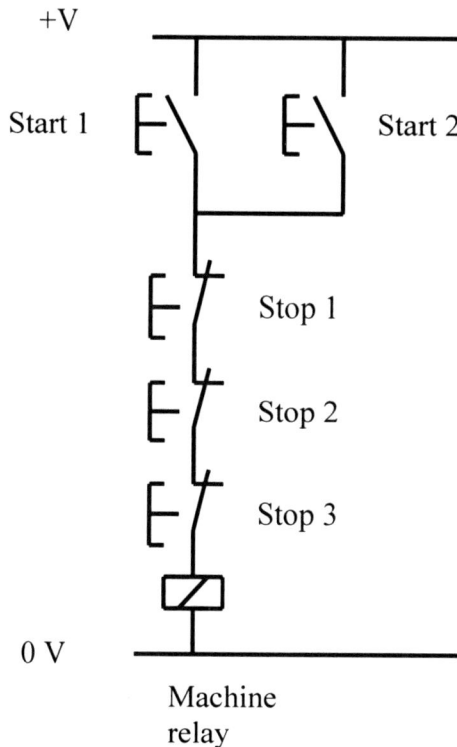

Fig 2.1 Hard-Wire Conditional Logic Example

The two *start* switches are connected in parallel. Current will flow if one *or* the other *or* both are closed. The start switches are normally open. This means that the contacts are apart and no current flows when the switches are in their normal (or *unoperated* or *rest*) state.

The three *stop* switches are connected in series. Current can only flow if the first *and* the second *and* the third are closed. The *stop* switches are normally closed. This means that the contacts are connected and current can flow when the switches are in their normal state.

The relay is a switch with multiple contacts that is operated when its coil is energised. The contacts are usually capable of carrying a larger current than push-button or limit switches. Large relays for motor starting are called contactors. The schematic diagram for a typical relay is shown in *Fig 2.2*.

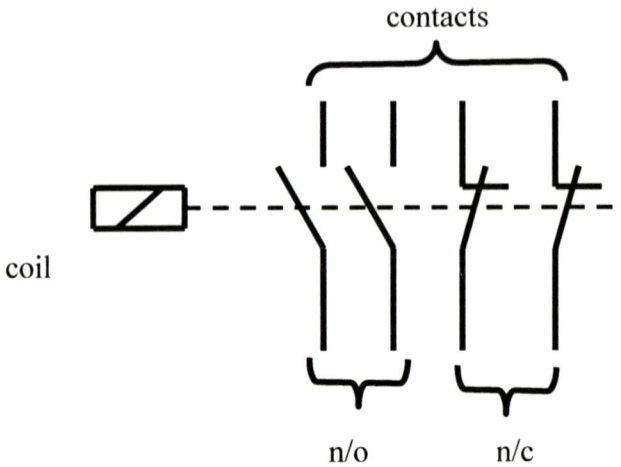

Fig 2.2 Relay

2.2 Ladder Diagrams

To realise the conditional logic statement from section 2.1 using ladder logic we connect the switches to a PLC as shown in *Fig 2.3*.

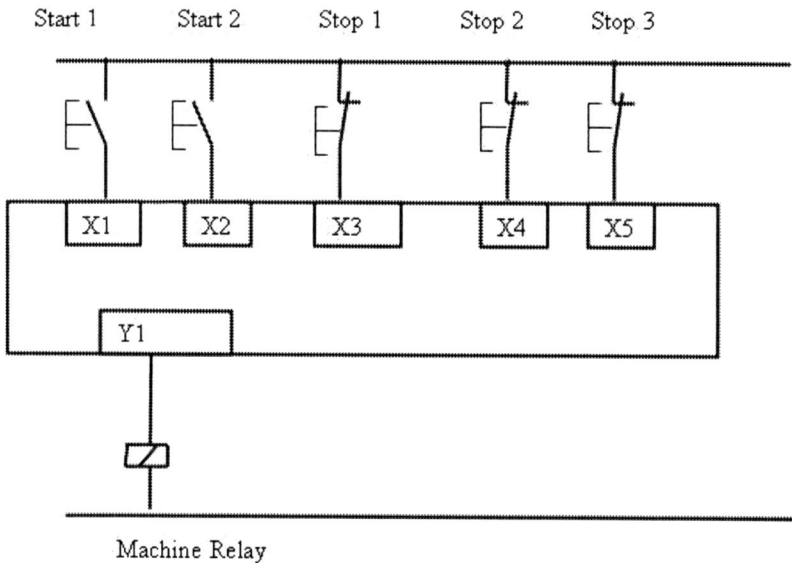

Fig 2.3 PLC Wiring Diagram

Fig 2.3 PLC Wiring Diagram

To avoid later confusion regarding the concept of normally open (n/o) and normally closed (n/c) it is worth looking again at Fig 2.3 and remembering that the plc scans each input and asks "Is it on or is it off?" The five switches shown are external devices and the PLC knows nothing about them. As far as the PLC is concerned, at the moment, inputs X1 and X2 are off and X3, X4 and X5 are on.

I have written the ladder logic using the TriLogi software. (For details of entering program elements see the Appendix)

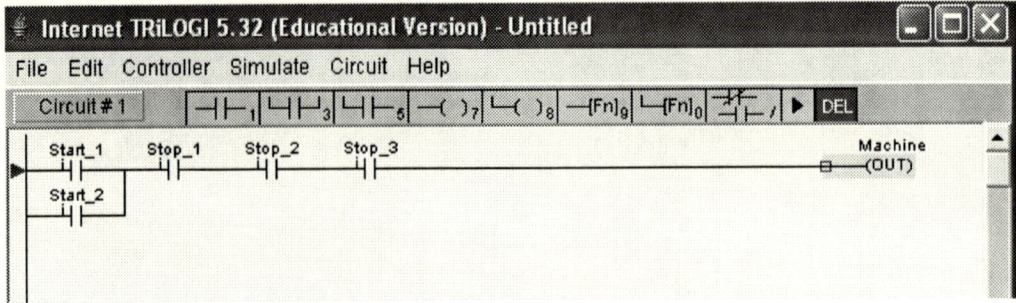

Fig 2.4 PLC Ladder Diagram

It can be seen from the Fig 2.3 and Fig 2.4 that the output *machine* will not be energised until one of the inputs *Start 1* or *Start 2* is switched on.Pushing any of the three *Stop* switches will turn off the input and so de-energise the output. It is normal practice to use normally closed push-button switches for *stop* buttons so that a failure of control voltage supply has the same effect as the pressing of the *stop* button.

2.3 Normally closed contacts

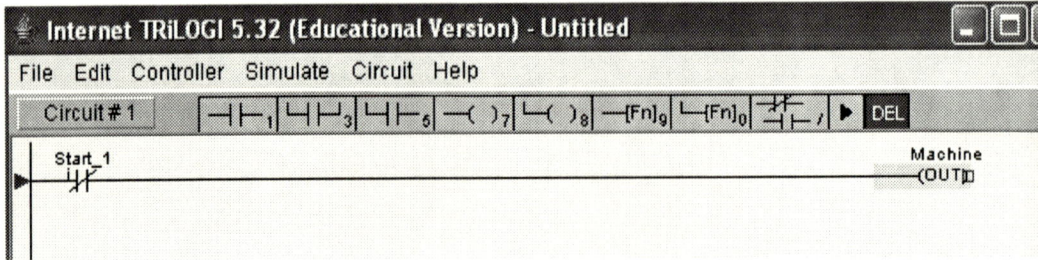

Fig 2.5 Normally closed contact.

The contact Start 1 in Fig 2.5 will be closed when the input is switched off and so the output Machine will be switched on. Switching on the input opens the contact and switches the output off. Remember that the nature (n/o or n/c) of the external switch that turns the input on, has no effect on the ladder logic.

2.4 Outputs and latches

Output states (*on* or *off*) can be used in programs as conditions for other actions.

Fig 2.6 is the wiring diagram for the program shown in *Fig 2.7*.

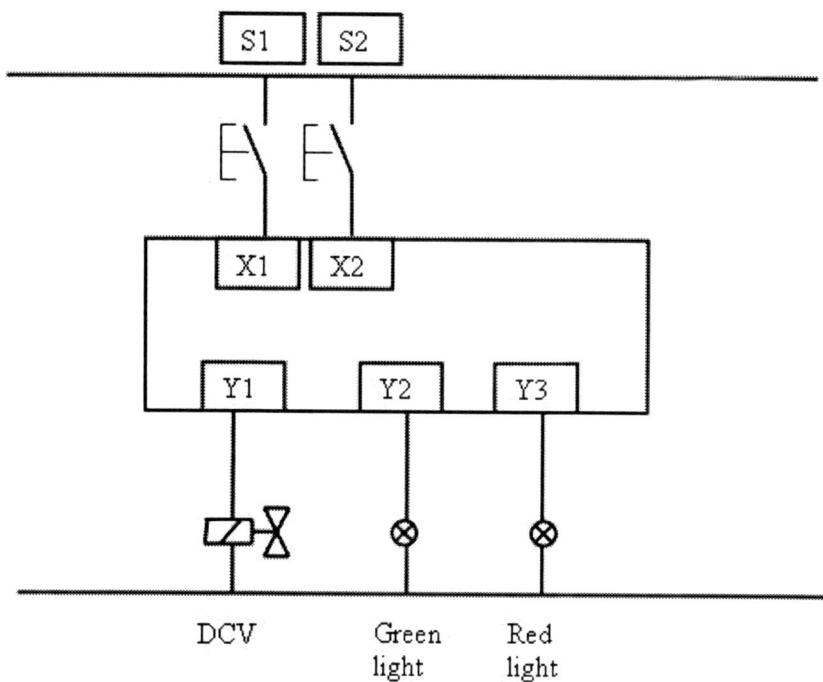

Fig 2.6

Fig 2.6

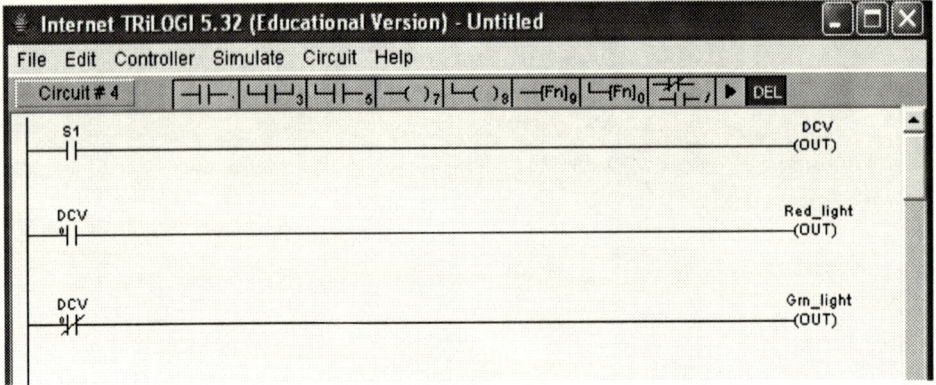

File Edit Controller Simulate Circuit Help

Circuit # 4   ⊣⊢ ⊣⊢₃ ⊣⊢₅ ⊣( )₇ ⊣( )₈ ⊣[Fn]₉ ⊣[Fn]₀ ⊣┼⊢ ▶ DEL

```
    S1                                                    DCV
    ⊣⊢─────────────────────────────────────────────────(OUT)


    DCV                                                   Red_light
    ⊣⊢─────────────────────────────────────────────────(OUT)


    DCV                                                   Grn_light
    ⊣/⊢────────────────────────────────────────────────(OUT)
```

Fig 2.7

Switching on the input S1 switches on the output DCV which in turn switches on the red light. When the output DCV is off the green light is on.

Example 2.1

Write a PLC program to implement the conditional logic statements (a), (b) and (c) below.

(a) A PLC output is to switch on if any of three inputs is switched on.

(b) A PLC output is to switch on if any *one* of three inputs is switched on but not two or more.

(c) A PLC output is to switch on if any two outputs are switched on, but not the third.

20

Solution

(a)

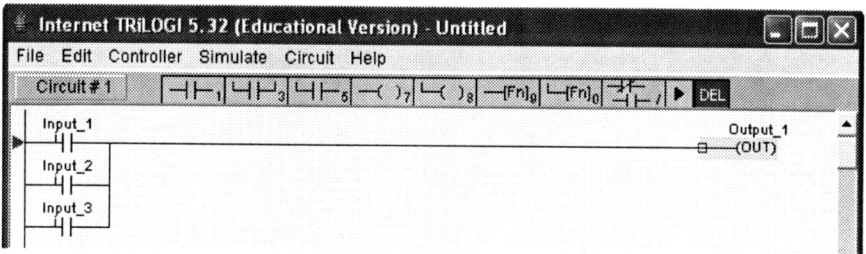

Fig 2.8

(b)

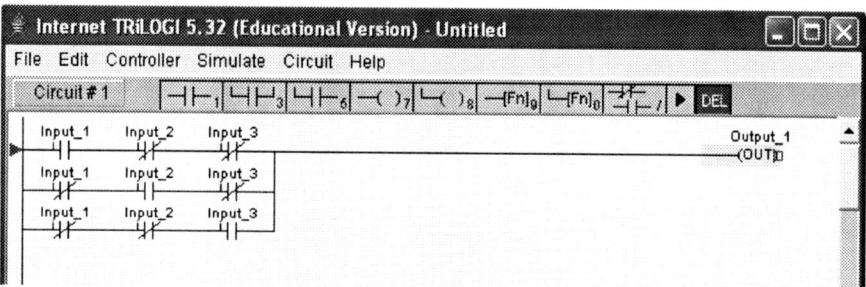

Fig 2.9

This program can be read:

"The output switches on if Input 1 is on AND the other two are off, OR input 2 is on AND the other two are off, OR input three is on AND the other two are off."

(c)

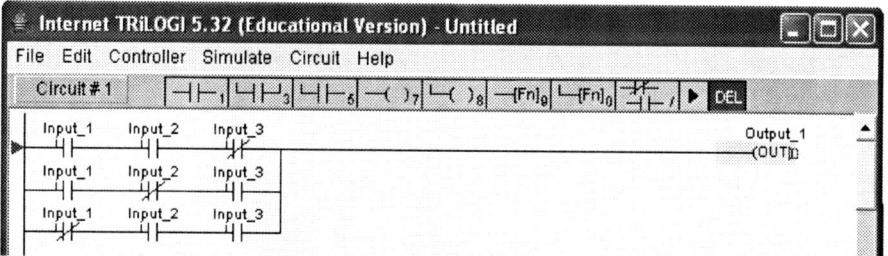

Fig 2.10

This program is similar to (b) above.

The push button and limit switches most commonly used in industrial automation are the *momentary contact* type. A spring action reverts the switch to the normal state as soon as the button or roller is released. These are obviously not the same as the *self- latching* switches used, for example, in domestic circuits.

The fact that the majority of control switches are not self-latching is not as inconvenient as it sounds. We can easily program in a latch in the ladder diagram.

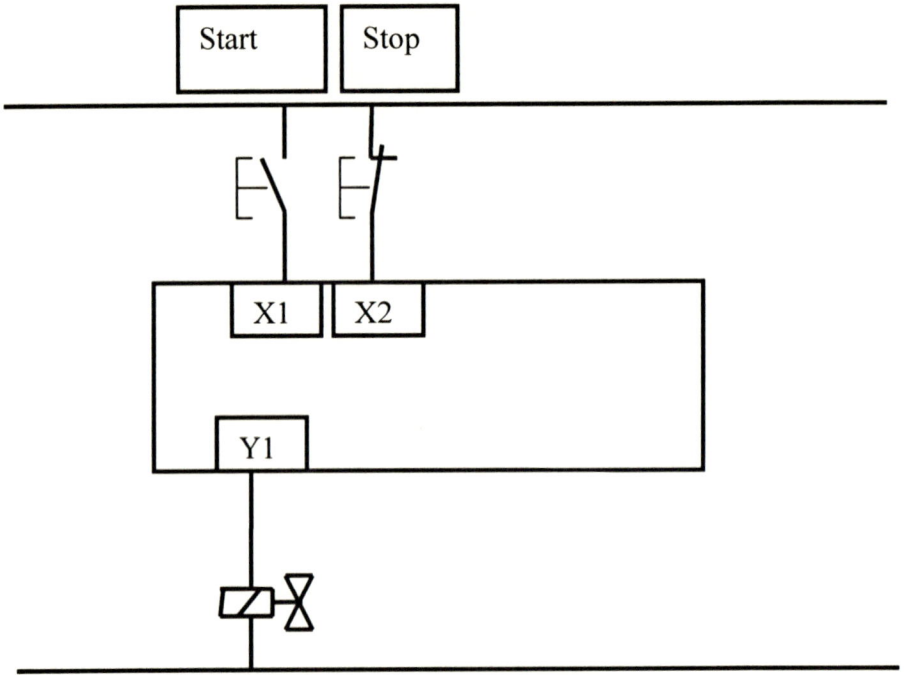

Fig 2.11

When the *start* push button switch in *Fig 2.11* is pressed, the output Y1 is to switch on and stay on until the *stop* button is pressed.

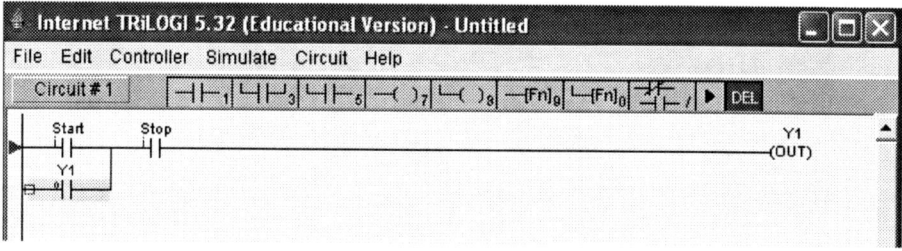

Fig 2.12

When the output Y1 is energised we use a normally open contact of it in parallel with the *start* button to hold (or latch) it on. The output can only be de-energised by the pressing of the *stop* button. Note that we have used a normally closed switch as a stop button as explained in section 2.2.

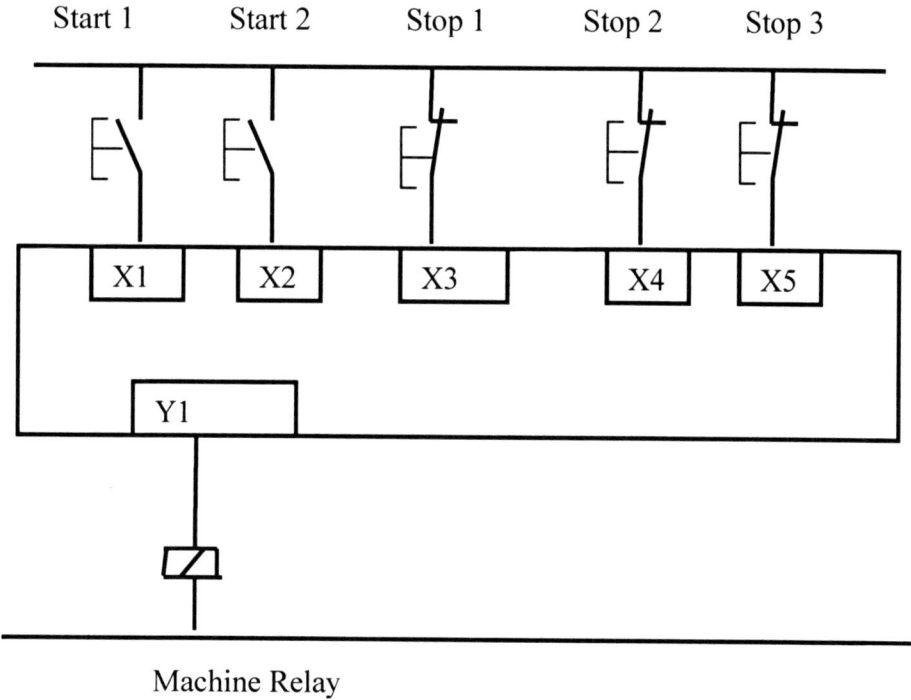

Fig 2.13

23

The latch concept can be extended to any number of start and stop switches. The output Y1 in *Fig 2.13* is to be switched on by X1 or X2 and is to stay on until any of the inputs X3, X4 or X5 is switched off. The required ladder diagram is shown in *Fig 2.14* below.

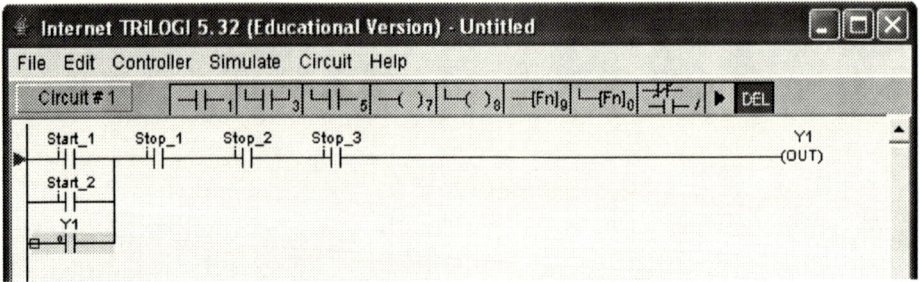

Fig 2.14

2.5 Internal relays

These have the same properties as outputs but they only exist in software. They have many uses. *Fig 2.15* shows an internal relay being used to implement the logic function NAND. This is the inverse of the result of X1 AND X2. We will be making extensive use if internal relays later in the book.

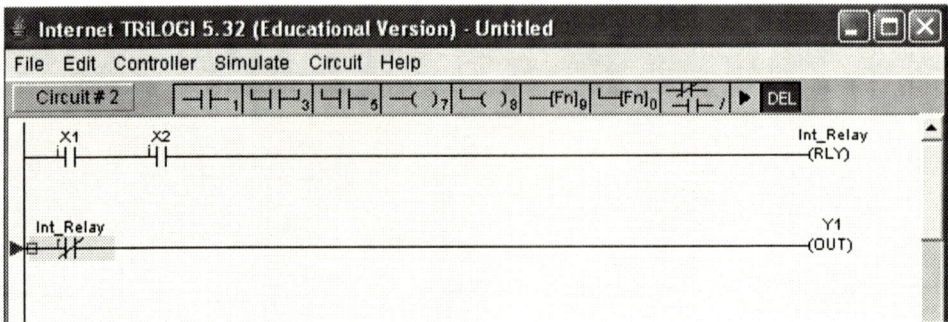

Fig 2.15 Use of internal relay

Note: Most PLCs include a function called a *Set and Reset* or a *flip-flop* which latches and delatches an output or an internal relay.

Throughout this book I use the latch as described in section 2.4, because of the visual resemblance of the ladder rung to the equivalent hard-wire circuit, in which a relay coil is latched on by a normally-open contact connected in parallel with the start button.

2.6 Timers

The delay-on timer introduces a delay between the start of one event and the start of another.

For example, when a start push button is pressed, the pneumatic cylinder shown in *Fig 2.16* extends, remains extended for 5 seconds and then returns. Draw the PLC wiring diagram and the appropriate ladder logic.

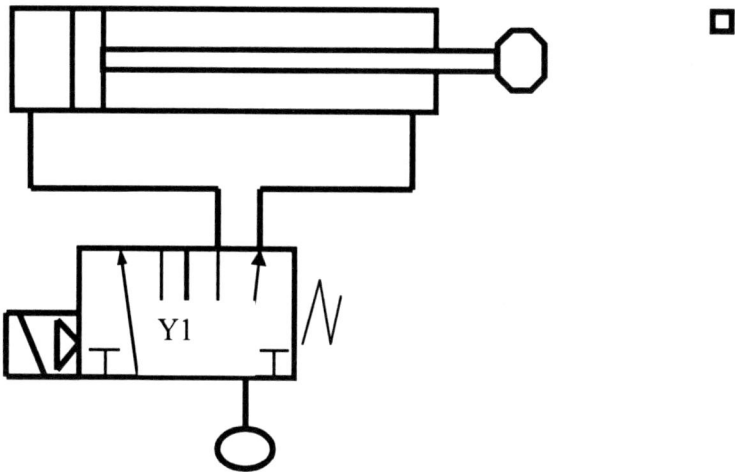

Fig 2.16

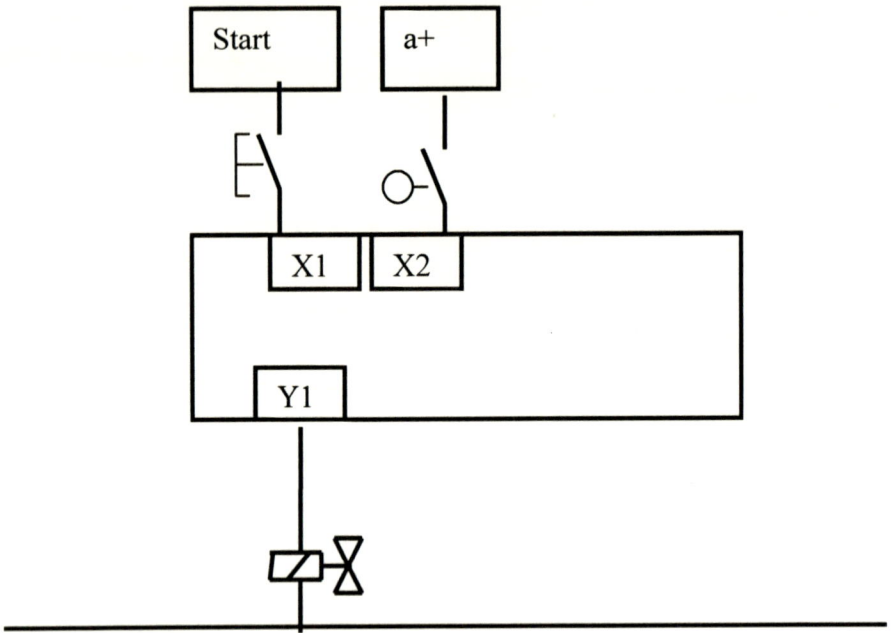

Fig 2.17

The *start* button and the end-of-stroke limit switch *a+* are the PLC inputs and the solenoid Y1 is the output. Any other components needed for the program can be created in software.

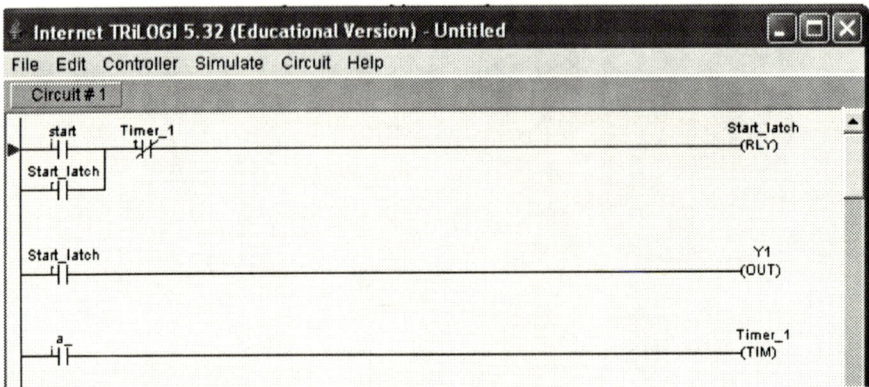

Fig 2.18

Pressing the *start* button latches on an internal relay called *start_latch*. The *start_latch* relay switches on the output Y1 which energises the solenoid, and the cylinder extends. The cylinder rod closes the limit switch a+ which starts the timer in software. When the timer set value time has elapsed the normally-closed contact *Timer_1* in the first line of the program de-energises the *Start_latch* relay and the cylinder returns.

The timer set value in the TRiLOGI software is in units of 0.1 s. For a 5 s delay a value of 50 is entered in the drop-down menu. (More details are given in the appendix)

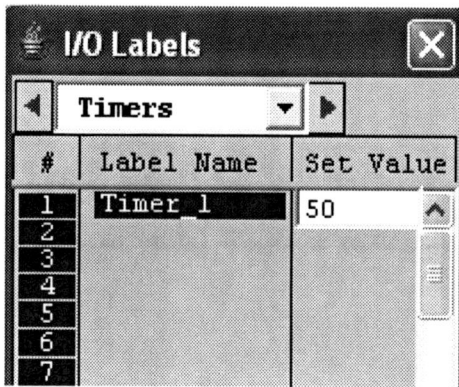

Fig 2.19 Setting Timer preset Value

We can do another example using the same hardware with the addition of an alarm as a second output:

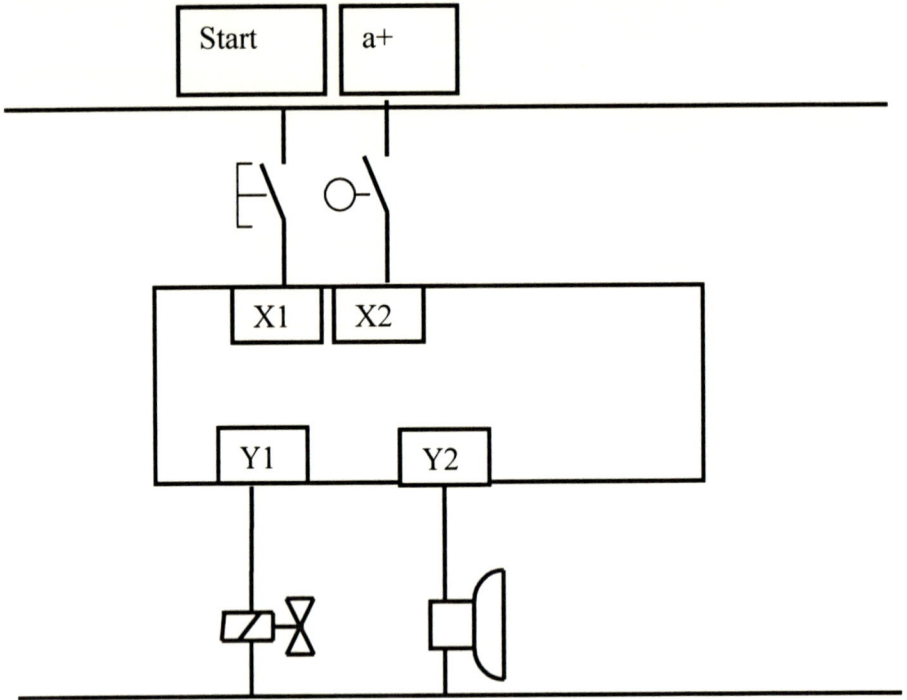

Fig 2.20

When the *start* push button is pressed and released there is a 5 s delay before the cylinder extends and returns. An alarm sounds during the 5 s delay.

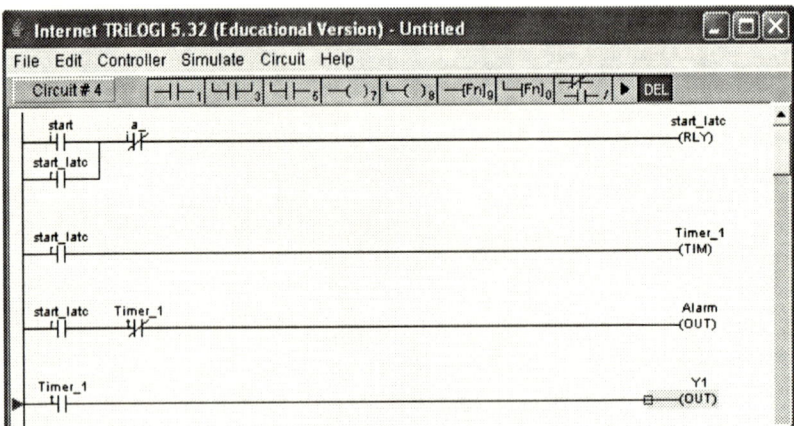

Fig 2.21

28

When the *start* button is pressed the *start_latch* relay is energised. The *Start_latch* relay switches on the timer input and the alarm. When the timer set value has elapsed the alarm switches off and the solenoid Y1 is energised. When the cylinder is fully extended the limit switch a+ de-energises the *start_latch* relay which de-energises the solenoid and resets the timer.

The input to the delay-on timer must remain on for the duration of the timer set value otherwise the timer will not operate. If the signal to start the timer is only momentary then a latch is used to sustain it. When the input to the timer switches off, the timer contacts revert immediately to their normal states.

In some PLC models a timer function block can be is located in the centre of a rung as shown in *Fig 2.22*. When the timer set value has elapsed the timer output switches on allowing a software signal to energise an internal relay coil or an output. In this book all timer function blocks are located at the right hand side of the ladder diagram and their contacts, normally-open or normally closed, have the same label as the timer.

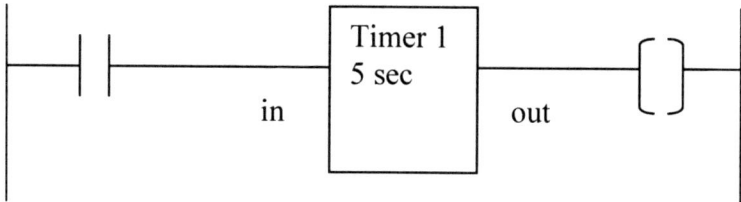

Fig 2.22

The delay-off timer causes a delay between its input switching off and its contacts reverting to their normal states. The delay-on timer is used throughout this book.

29

2.7 The Pulse Generator

Two counters can be combined to make a pulse generator. This is best illustrated by an example.

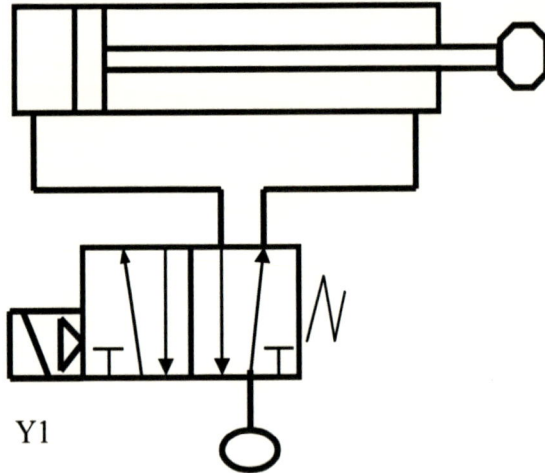

Y1

Fig 2.23

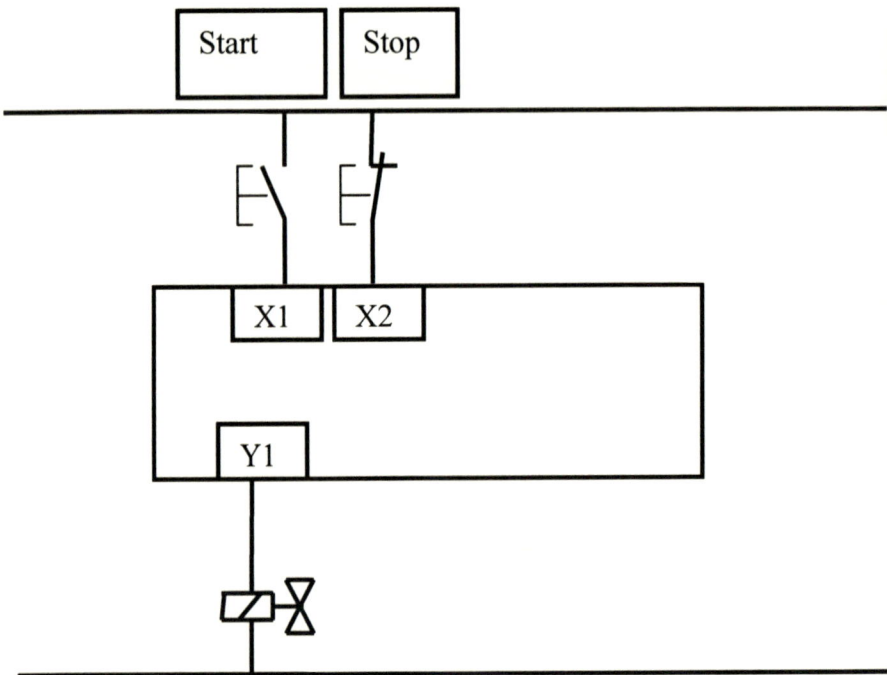

| Start | Stop |

X1   X2

Y1

Fig 2.24

When the *Start* button in *Fig 2.24* is pressed the cylinder in *Fig 2.23* oscillates, extending for 2 s and returning for 1 s until the *Stop* button is pressed.

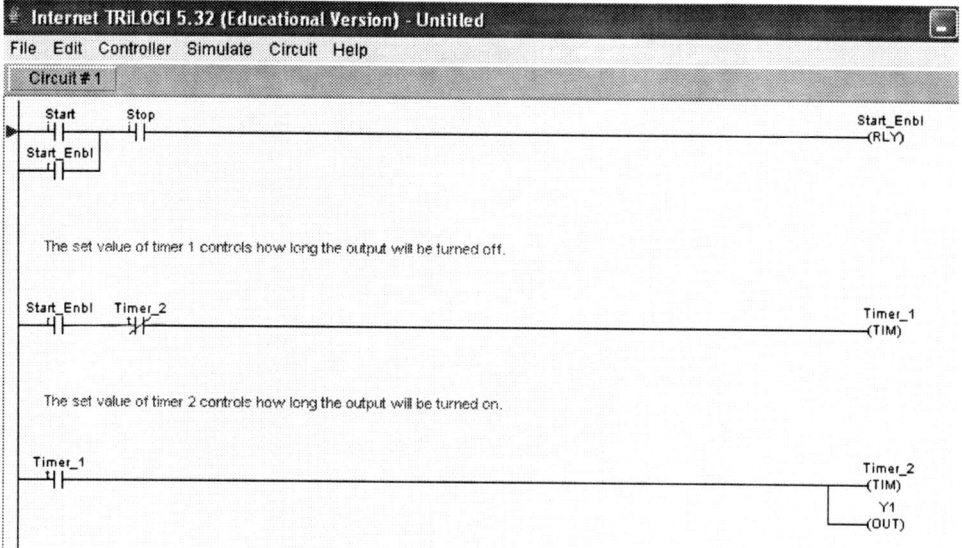

Fig 2.25 Pulse Generator

It can take a while to figure out how the pulse generator works but it is time very well spent. The flow chart in *Fig 2.26* should help.

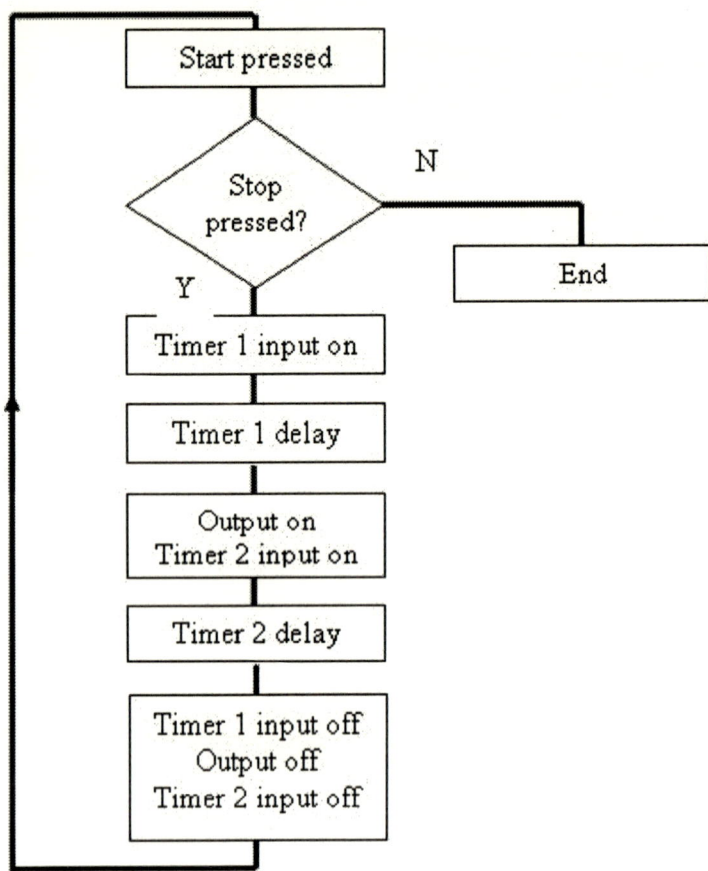

Fig 2.26 Pulse Generator Flowchart

Fig 2.26 Pulse Generator Flowchart

2.7 Counters

A counter allows a number of occurrences of input signals to be counted. The counter is set to a *preset number value* and when this value of input pulses has been received, it will operate its contacts. A second input or software coil is provided to reset the *current value* of the counter to zero.

Consider the cam shaft in *Fig 2.27*.

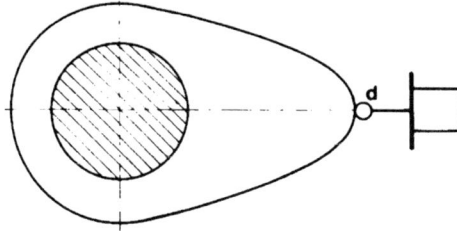

Fig 2.27

When a *start* button has been pressed the shaft is to make 10 revolutions and then stop. Pressing the start button also resets the counter. The PLC wiring diagram is shown in *Fig 2.28*.

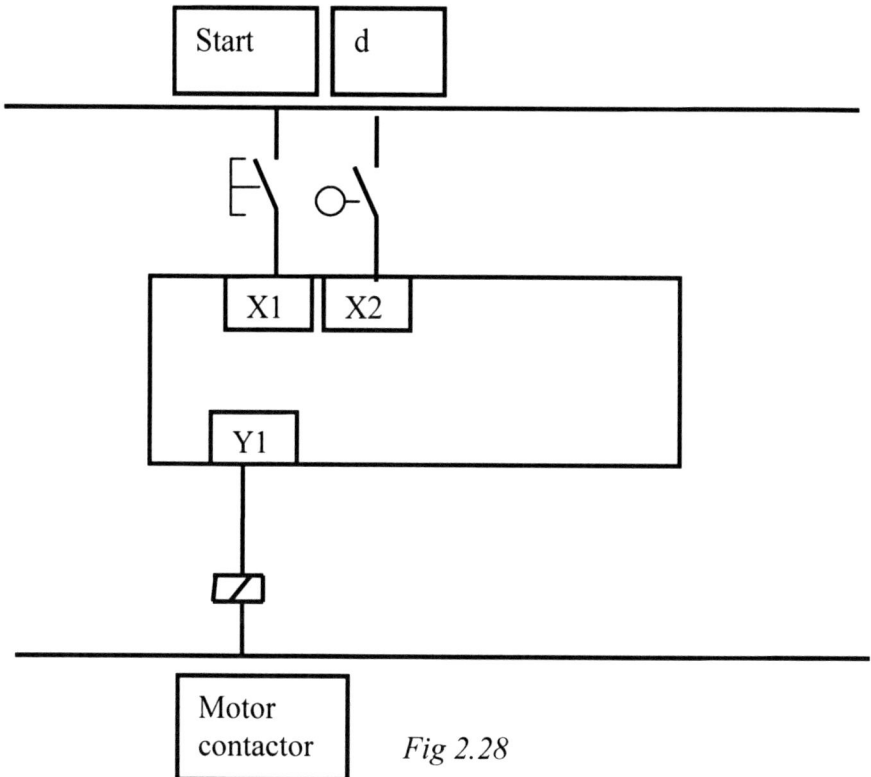

Fig 2.28

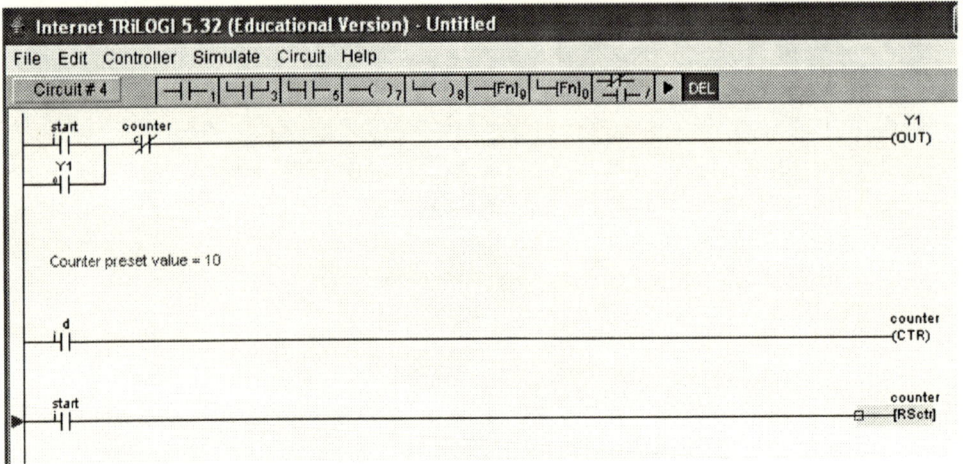

Fig 2.29

The pulse generator and counter can be combined as shown in this final example.

When a *start* push button is pressed and held down, an alarm sounds six times before a conveyor starts. Pressing the conveyor *stop* button also resets the counter. *Fig 2.30* and *Fig 2.31* show a solution to the problem.

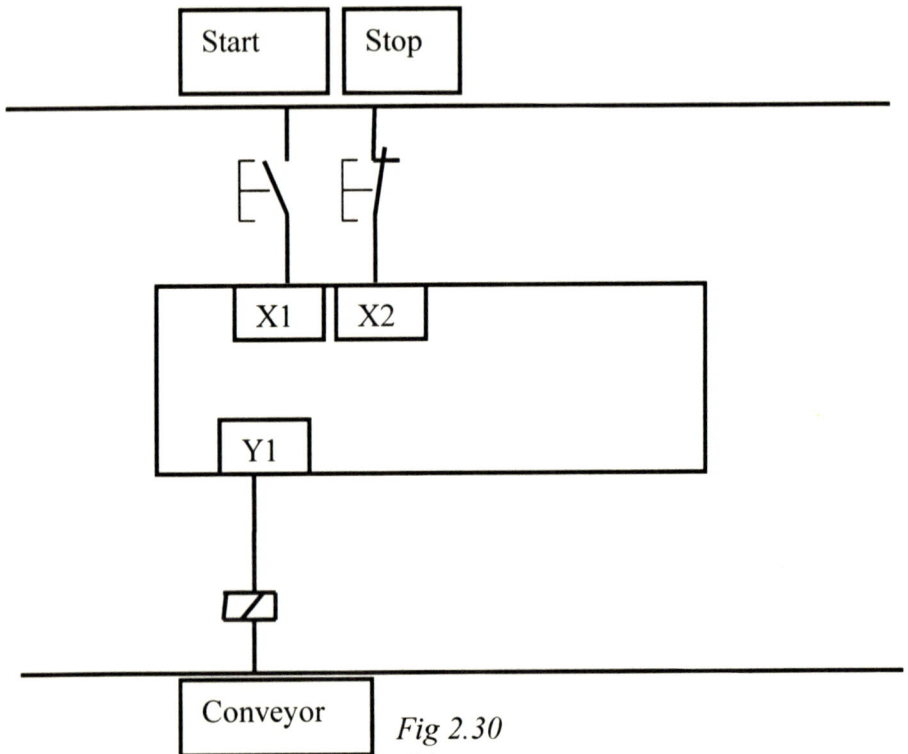

Fig 2.30

34

3. Which form of logic gate system is given by a ladder diagram with a rung having two normally open sets in parallel as shown?

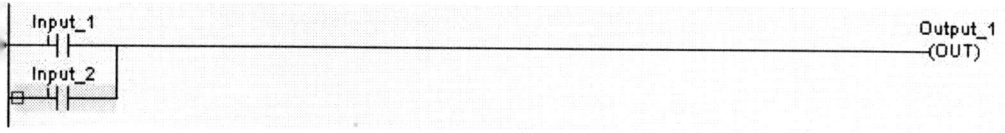

Fig 2.34

Answer:

- a. OR
- b. NOR
- c. AND
- d. NAND

4. Which form of logic gate system is given by a ladder diagram with a rung having two normally open sets of contacts in series as shown?

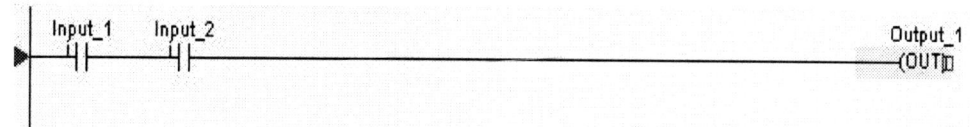

Fig 2.35

Answer:

- a. NOR
- b. NAND
- c. OR
- d. AND

The PLC diagram *Fig 2.36* applies to questions 5-10.

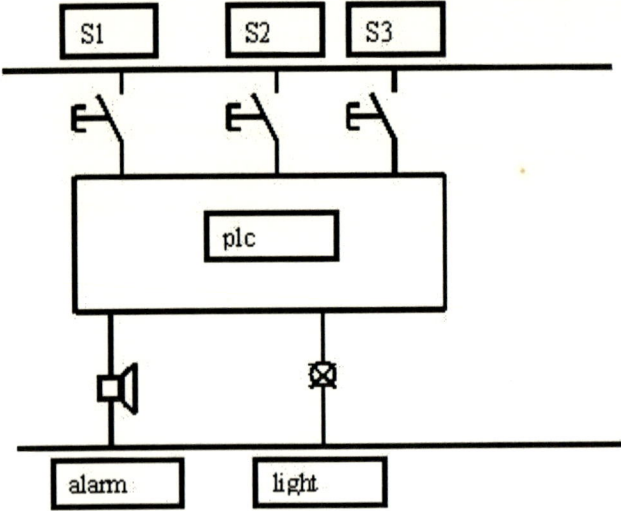

Fig 2.36

5.

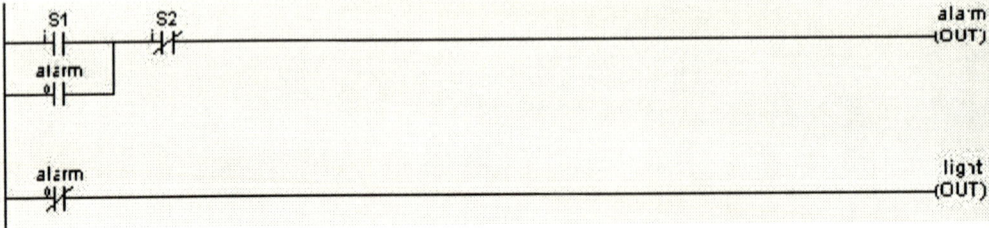

Fig 2.37

Pick the one <u>correct</u> statement below regarding the ladder diagram *Fig 2.37*

Answer:
  ○ a. Closing switch S1 switches off the light

  ○ b. When the alarm is on, so is the light.

  ○ c. Closing S1 and S2 switches on the alarm.

  ○ d. Closing switch S2 switches on the alarm

6.

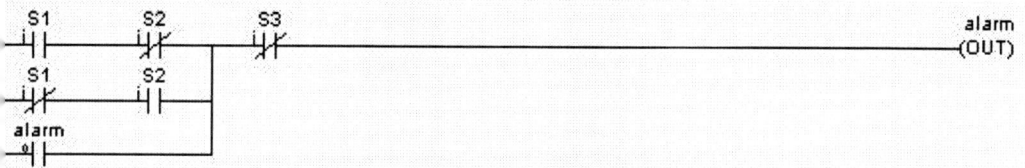

Fig 2.38

Pick the one <u>incorrect</u> statement below regarding the ladder diagram *Fig 2.38*.

Answer: ○ a. When the alarm is switched on it keeps going until S3 switches on.

○ b. S1 and S2 form an exclusive or (XOR) function

○ c. The alarm is switched on when S1 or S2 or S3 is on.

○ d. The alarm is started when S1 or S2 is switched on but not both together

7.

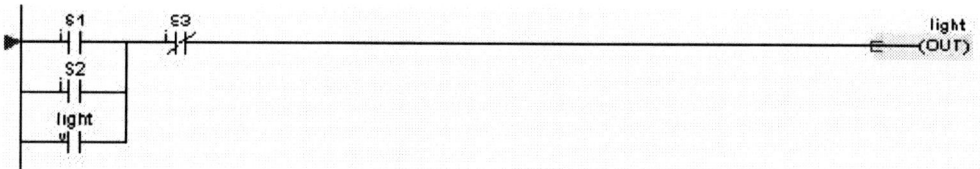

Fig 2.39

Pick the one <u>correct</u> statement below regarding the ladder diagram *Fig 2.39*

Answer: ○ a. If switch S3 is closed the light will be on

○ b. When the light is on it stays on until S1 or S2 is pressed.

○ c. S1 and S2 form an exclusive or (XOR) function

○ d. The light is switched on by pressing S1 or S2

39

8.

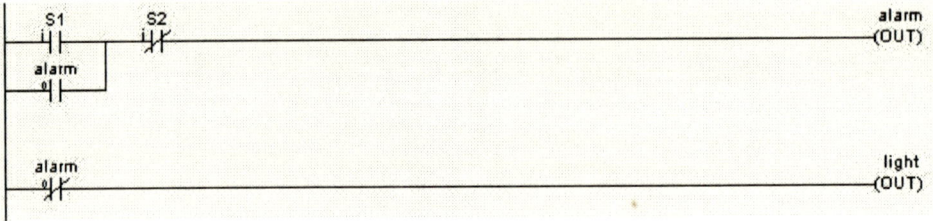

Fig 2.40

Pick the one <u>incorrect</u> statement below regarding the ladder diagram *Fig 2.40*

Answer: ○   a. S2 can be used to switch off the light before the timer delay is complete.

○   b. The light output is latched on.

○   c. When the timer delay is finished the light comes on.

○   d. The light remains on for a time equal to the timer delay setting.

9.

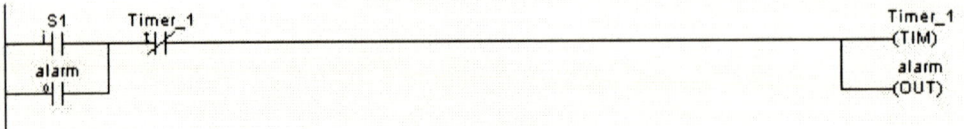

Fig 2.41

Pick the one <u>correct</u> statement below regarding the ladder diagram *Fig 2.41*

Answer:

    ◌   a. When switch S1is pressed and released there is a delay equal to the timer setting before the alarm sounds.

    ◌   b. If S1 is to latch on immediately it has been pressed, a normally-open timer contact should be connected in parallel with it.

    ◌   c. When switch S1is pressed and released the alarm sounds for a time equal to the timer setting.

    ◌   d. The normally-closed timer contact prevents the alarm sounding or the timer being energised.

10.

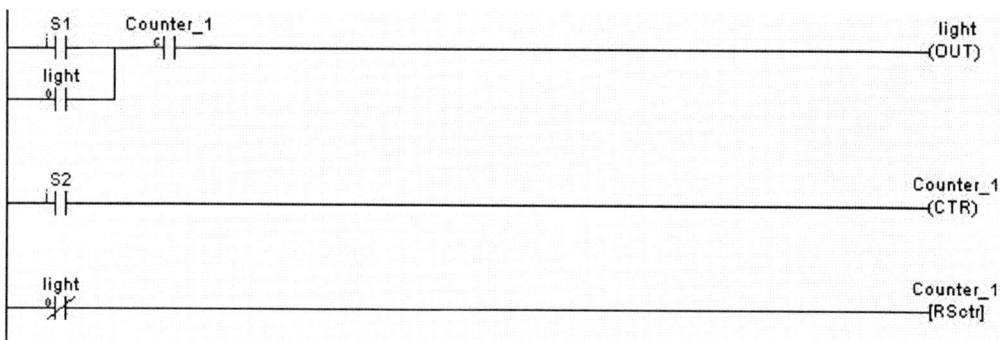

Fig 2.42

In the ladder diagram *Fig 2.42* the counter preset = 5. Choose the one <u>incorrect</u> statement below

Answer:

    ◌   a. When S1 is momentarily pressed the light comes on and stays on.

    ◌   b. When the light goes out the counter is reset.

    ◌   c. When switch S1 is pressed the light will come on until S2 is pressed.

    ◌   d. When switch S1 is pressed the light will come on until S2 is pressed 5 times

11. A PLC is to be used to control a flood light. When a sensor with a normally open contact detects movement the light is to switch on for 10 seconds and then switch off.

Draw the necessary PLC wiring diagram and the ladder logic to operate the system as designed.

12. A PLC is to be used to control the drive for a car window. When a momentary contact switch switch is pressed the window starts to open. If the switch is closed for more than 1 second, the window contunues opening until fully open. A second switch does the same thing to close the window. Limit switches are provided to detect the window fully open or fully closed positions. Draw the necessary PLC wiring diagram and the ladder logic to operate the system as designed.

13. A PLC is used to control a conveyor system. A sensor with a normally open contact sees items passing on the conveyor. When 10 items have passed, the conveyor stops, a cylinder extends and retracts and the conveyor runs again until another 10 items have passed. Draw the necessary PLC wiring diagram and the ladder logic to operate the system as designed.

Chapter 3
Sequential Programming

3.1 Introduction

Most machine operations are sequential in nature so it is necessary for the PLC to switch outputs depending not only on the input combinations but also on the current stage in the sequence. An output operating at the wrong time could cause damage or injury so the correct programming technique is critical.

3.2 A simple automation sequence

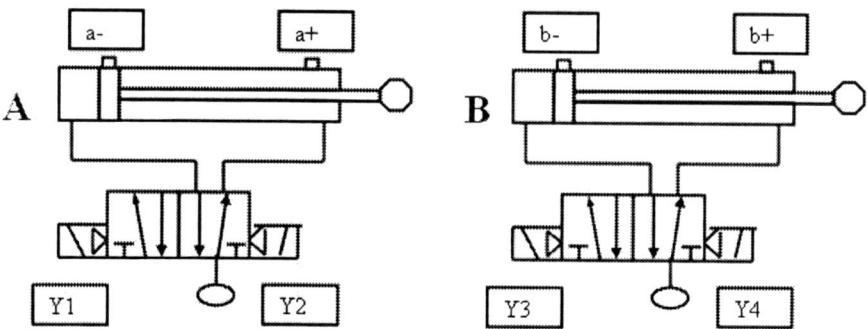

Fig 3.1

The two cylinders A and B in Fig 3.1 are to go through the sequence

A+ B+ A- B-

Reed switches a-, a+, b- and b+ have been fitted to detect the magnetised cylinder pistons through the aluminium cylinder bodies. The 5 port 2 position directional control valves (5/2 DCVs) are double-solenoid operated.

The PLC wiring diagram is shown in *Fig 3.2*.

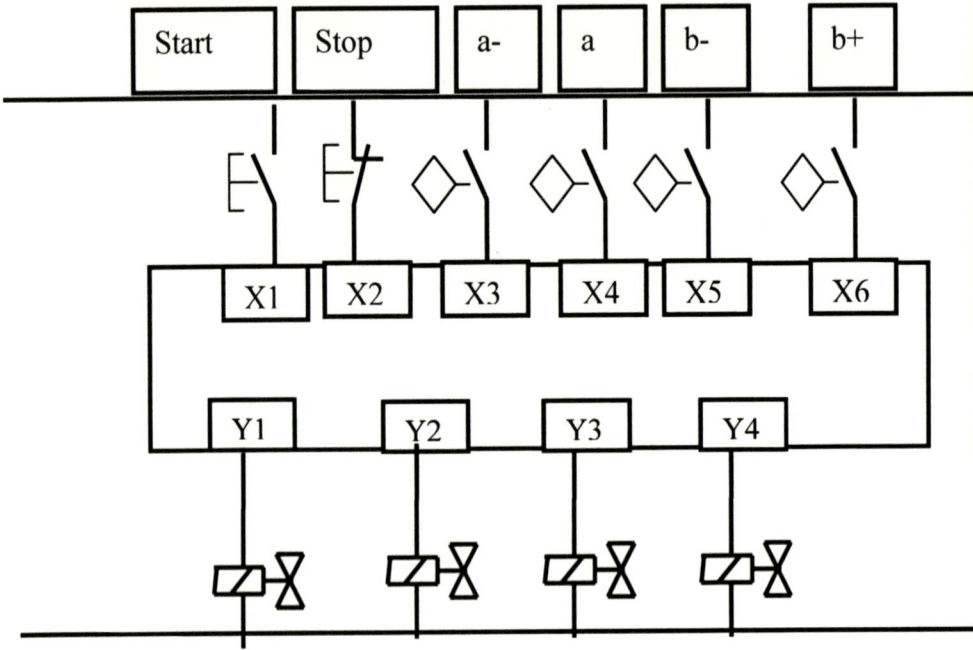

Fig 3.2

Each event in a sequence is started by the completion of the previous event. When reed switch a+ ,for example, closes, it signals the end of event A+ (the extension of cylinder A) and the beginning of B+ (the extension of cylinder B).

We will write the program line-by-line on this basis.

Fig 3.3

Pressing the *Start* button causes the cycle to execute once. The reed switch *b-* is not required neither is the *Stop* button

Latching on the *Start* button with an internal relay and encorporating the reed switch *b-* causes the cycle to repeat until the *Stop* button is pressed. This is shown in *Fig 3.4*.

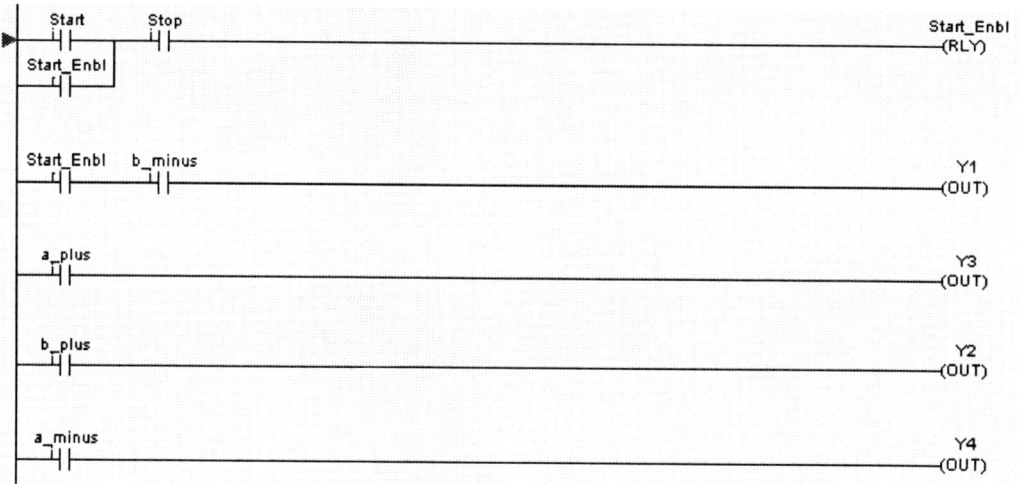

Fig 3.4

It all seems pretty straightforward so far doesn't it? Let's try another sequence using the same hardware.

A+ B+ B- A-

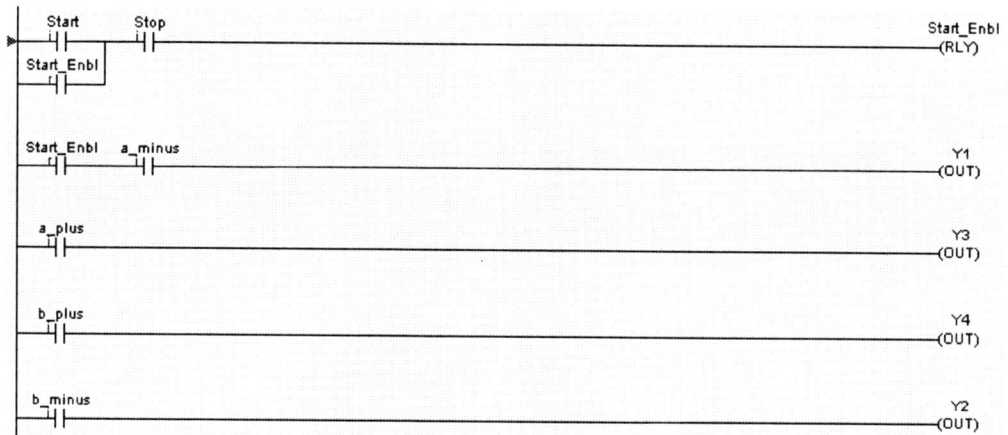

Fig 3.5

I have entered the ladder logic in a similar way to the first sequence. Everything works fine until we get to the third rung of the program where the reed switch *b+* is supposed to energise solenoid Y4 to cause cylinder B to return. At this point both cylinders are extended as shown in *Fig 3.6*.

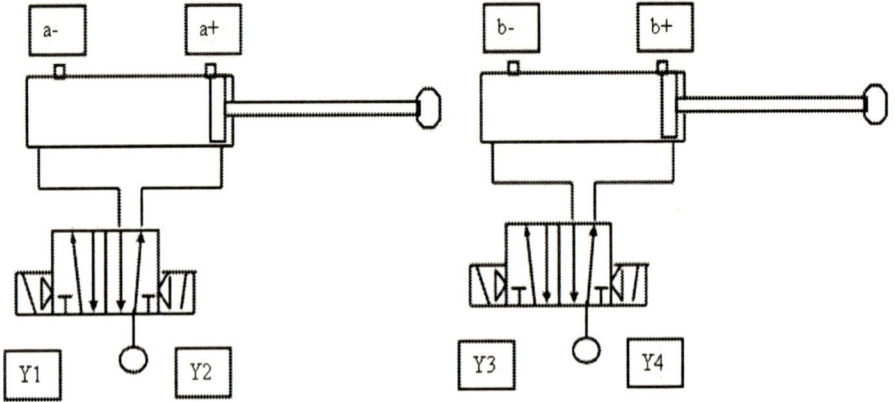

Fig 3.6

The fact that cylinder A is also extended means that reed switch *a+* is closed and therefore solenoid Y3 is energised, cylinder B therefore cannot return. This situation is called a *trapped* signal. It is characterised by having both solenoids of a double solenoid directional control valve simultaneously energised and it prevents us from programming many circuits in a simple sequential fashion.

The realisation of this sequence is even more difficult using single-solenoid, spring return directional control valves because latching is required and we are still only dealing with a two cylinder problem.

Trapped signals also occur in pneumatics and in electro-pneumatics and various methods are employed to get over the problem. The best known of these is the cascade system but it is only of practical use in simple systems.

3.3 Evolution of the Sequential Function Chart

As PLC sequences became more complex during the 1970s, the need grew for a universal programming method that would standardise PLC programs and also solve commonly encountered problems such as trapped signals.

In 1975 a working group, drawn from the Association Française pour la Cybernétique Economique et Technique, developed GRAFCET (GRAphe Foncionnel de Commande, Etapes, Transitions) which has since formed the basis of the Sequential Function Chart method of programming.

In the early 1990s the International Electro-technical Commission (IEC) a sister organisation to the International Standards Organisation published the IEC61131 standard, part 3 of which deals with programmable languages PLC software structure, languages and program execution. The standard identifies 5 distinct programming methods including ladder logic and the sequential function chart.

Until the IEC 61131-3 standard was published in March 1993, there was no suitable standard that defined the way control systems such as PLCs could be programmed.

Ladder Programming has become as one of the most popular graphical languages for programming PLCs mainly because of its ressemblance to hard-wire control circuits. Unfortunately its suitability for building complex sequences is limited.

The Sequential Function Chart (SFC) is an extremely effective graphical language for expressing the high level sequential parts of a control program. The best of both worlds approach is to plan the program using SFC and then to translate it into ladder logic, it is this approach that we will take to solving automation problems in the remainder of this book.

3.4 Programming Using the Sequential Function Chart

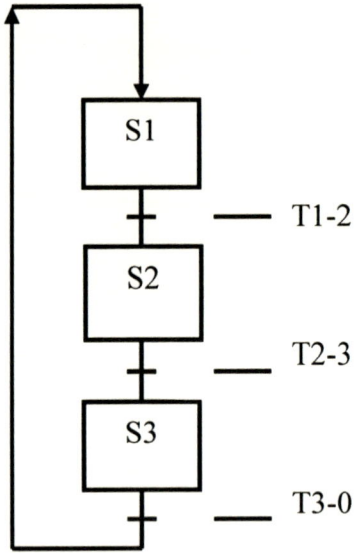

Fig 3.7 Sequential Function Chart

The system passes through successive *states* during which *events* take place.

The states are linked by *transitions* which provide the bridge from one state to the next.

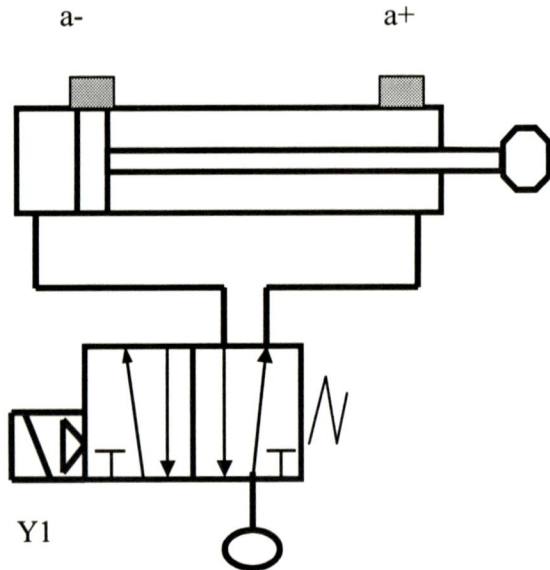

Fig 3.8

Example 3.1

When a start push button is pressed the cylinder in *Fig 3.8* is to extend and retract, repeating until a stop button is pressed. When the system is powered up and ready a green light Y2 is on and while the cylinder is oscillating, a red light Y3 is on.

We draw a PLC wiring diagram as usual.

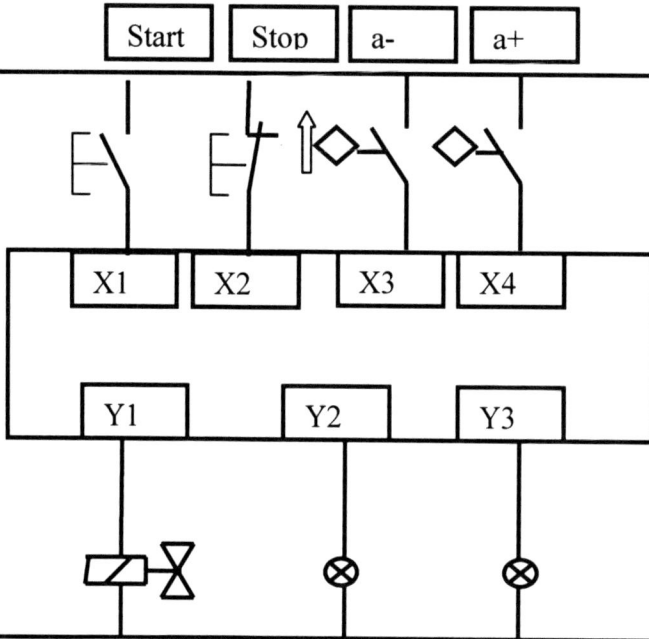

Fig 3.9

Now we complete the sequential function chart (SFC)

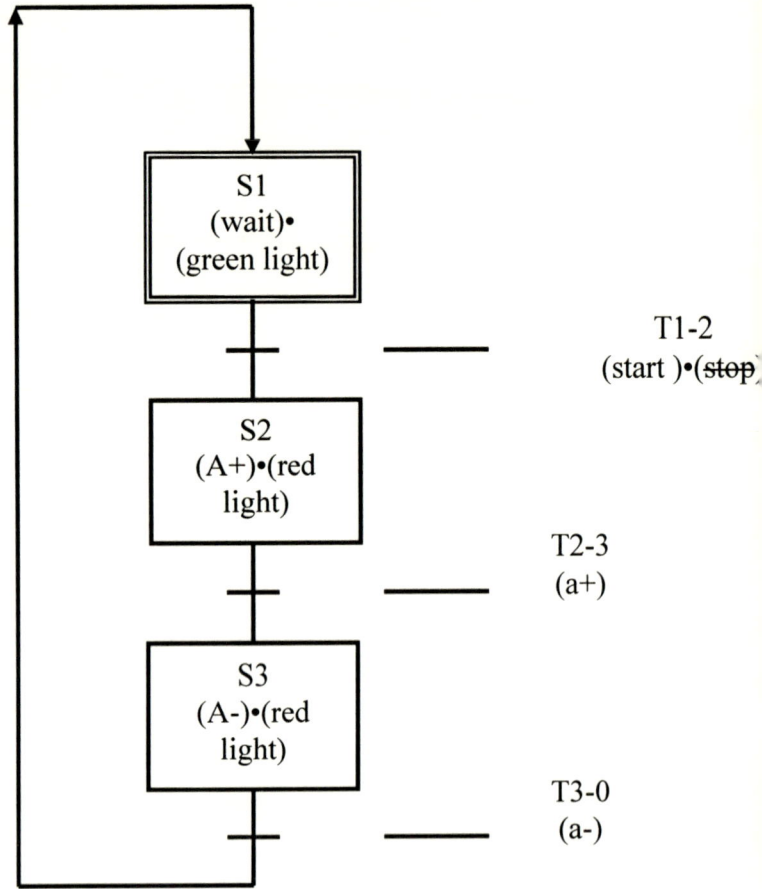

Fig 3.10 Completed SFC

States: S1, S2 and S3 are respectively States 1, 2 and 3. The event or events occurring during each state are written in brackets in the state box.

e.g.   S2 (A+)•(red light)   means that when state 2 is active cylinder A extends and (the dot • means *and*) the red light is on.

Transitions: The transitions represent the changeover from one state to another. The origin and destination states are indicated by the transition label. The input or timer that activates the transition is written in brackets under the transition title. The label (start)•(stop)

50

under T1-2 means that the start button has been pressed and the stop button has not been pressed. In ladder logic this translates as "The *start enable* relay is latched on."

3.5 Entering the SFC program into the PLC

Some PLC models accept SFC programs directly but we will enter the program as ladder logic.

Each state and transition is assigned an internal relay. When the state or transition is active, its relay is turned on. We write down the states and transitions with their associated relays in an *assignment list*.

S1	R1
S2	R2
S3	R3
T1-2	R4
T2-3	R5
T3-0	R6

Table 3.1 Assignment List

Now we're ready to enter the program. We put in the ladder logic in the following order.

Initial Conditions
Transitions
States
Outputs
Timers
Counters

51

The ladder diagram is shown below broken into sections for clarity (*Fig 3.11* to *Fig 3.15*) . In the program these sections just follow one another in sequence.

Initial Conditions

The only initial condition for this program is the latching on of the *start* button by an internal relay. Low numbered relays are mostly taken up as labels for states and transitions so I picked R50 for this purpose and for clarity I have used the *Start_Enbl* label.

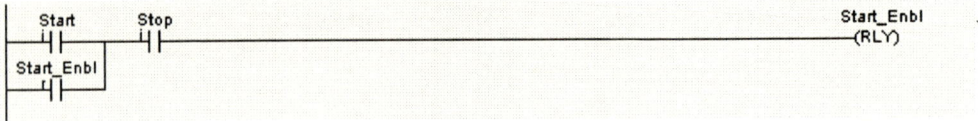

```
Start      Stop                                                    Start_Enbl
 ┤├        ┤├                                                       ─(RLY)
Start_Enbl
 ┤├
```

Fig 3.11

Transitions

There is a minimum of two conditions for a transition to occur. The relevant state must be active and the input (or timer contact) that indicates that the event is completed must be on. Table 3.1 shows the transition conditions for this program.

State	Transition Condition	Transition
S1	*Start_Enbl* relay	T 1-2
S2	Input *a+*	T 2-3
S3	Input *a-*	T 3-0

Table 3.2

52

Now we can enter the transitions as ladder logic.

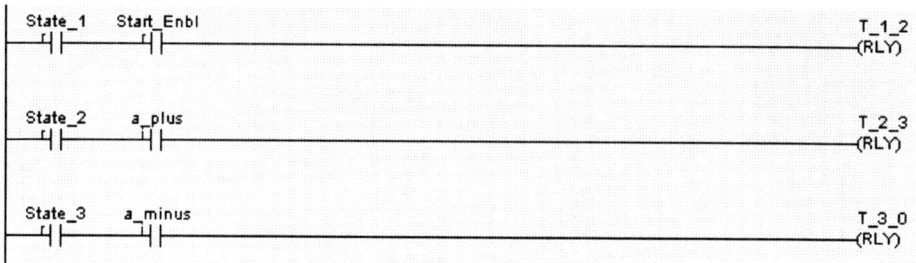

```
State_1    Start_Enbl                                          T_1_2
 ─┤ ├───────┤ ├──────────────────────────────────────────────(RLY)


State_2    a_plus                                              T_2_3
 ─┤ ├───────┤ ├──────────────────────────────────────────────(RLY)


State_3    a_minus                                             T_3_0
 ─┤ ├───────┤ ├──────────────────────────────────────────────(RLY)
```

Fig 3.12 Transitions

States

State 1 is the initial state. We need to make sure that it alone is active before the *start* button is pressed. To do this we use the fact that no internal relays are energised when the PLC is powered up. Stated verbally this reads "If state 2 relay is off and state 3 relay is off then state 1 relay switches on." *Fig 3.13* shows this entered as ladder logic.

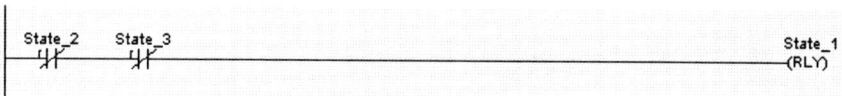

```
State_2     State_3                                           State_1
 ─┤/├────────┤/├──────────────────────────────────────────────(RLY)
```

Fig 3.13 Initial State

The remaining states are switched on and off by the relevant entry and exit transitions. The transition relays are often only briefly energised so the states are latched on. *Fig 3.14* shows the ladder logic for states 2 and 3.

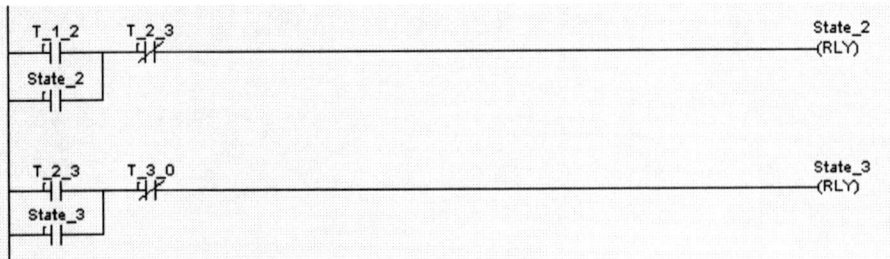

Fig 3.14 States 2 and 3

Outputs

Each output is switched on during the relevant state or states. This is shown in *Fig 3.15*.

Fig 3.15 Outputs

It seems like an awful lot of ladder logic for a simple program (and it is) but extra lines of software come free of charge and, as the programming tasks get more complex, the merits of sequential function chart programming quickly become apparent.

Example 3.2.

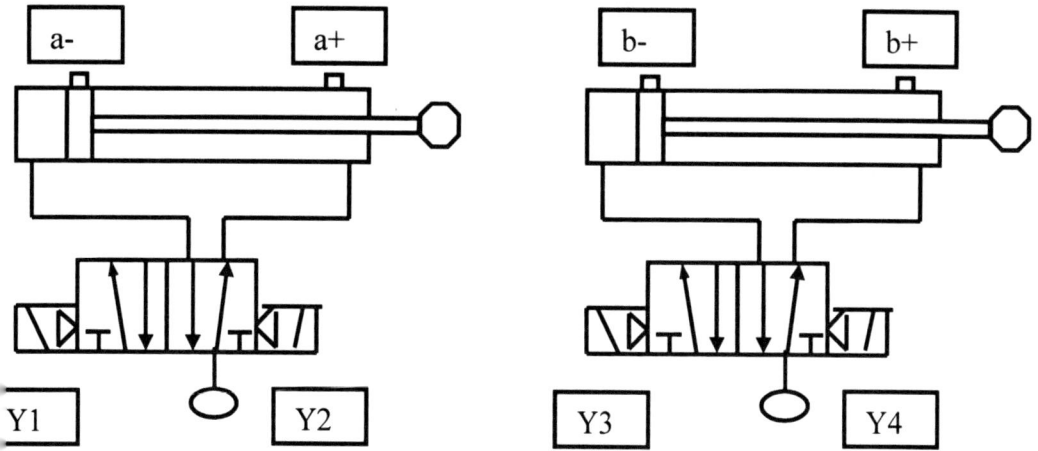

Fig 3.16

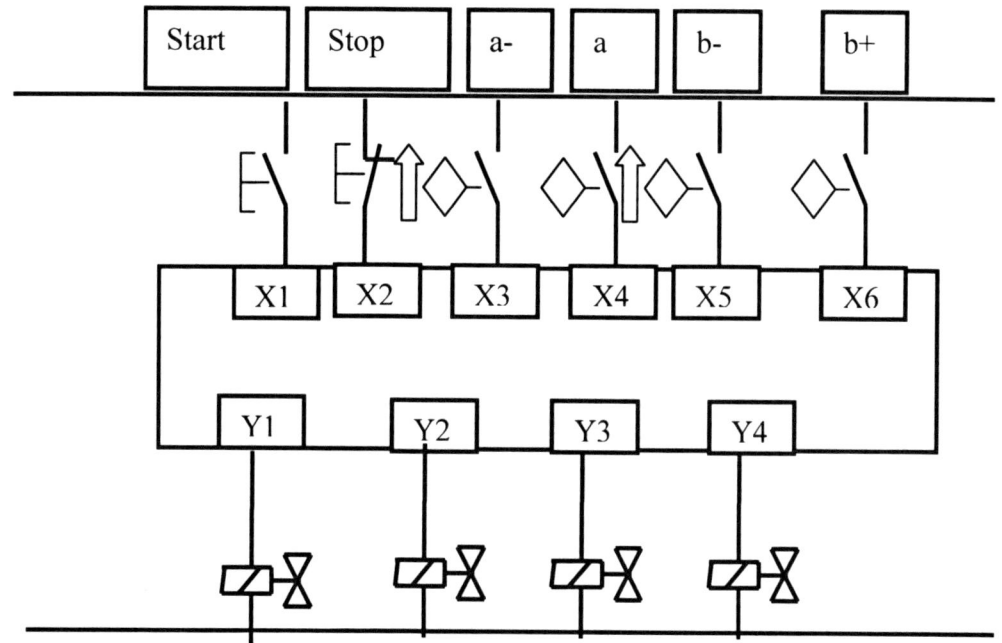

Fig 3.17

55

We'll use the SFC method to implement the sequence

A+ B+ A- B-

The SF chart is shown in *Fig 3.18* and the assignment list in *Table 3.3*.

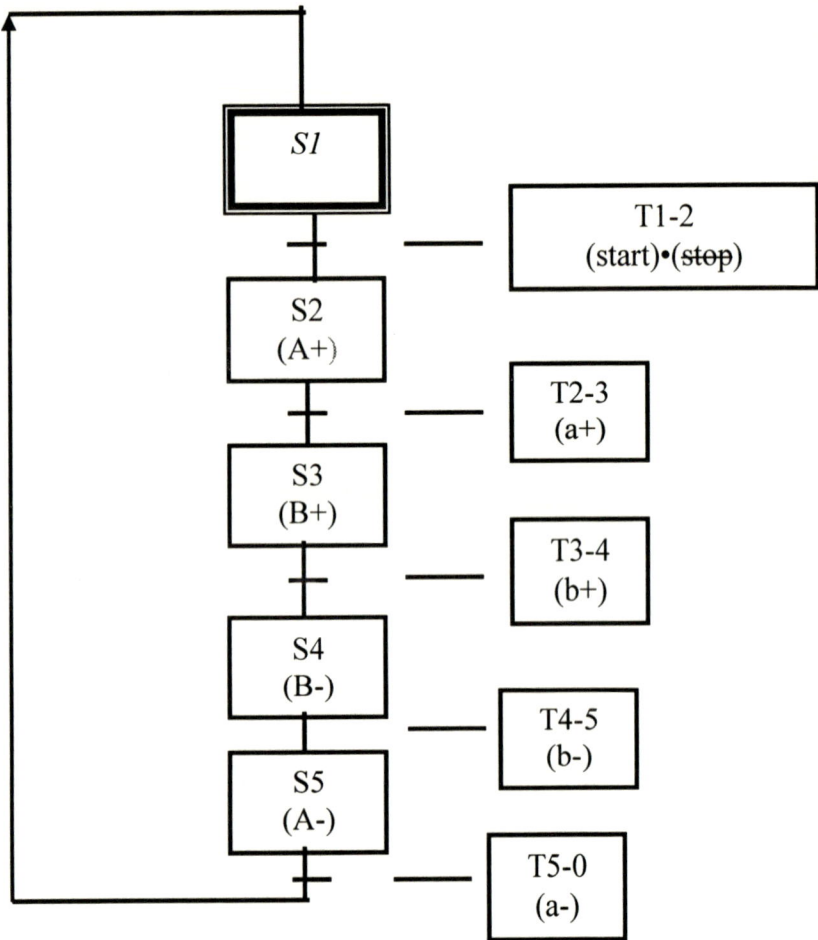

Fig 3.18

S1	R1
S2	R2
S3	R3
S4	R4
S5	R5
T1-2	R6
T2-3	R7
T3-4	R8
T4-5	R9
T5-0	R10

Table 3.3

Now we can enter the ladder logic, remember the order

Initial Conditions
Transitions
States
Outputs
Timers
Counters

Initial Conditions

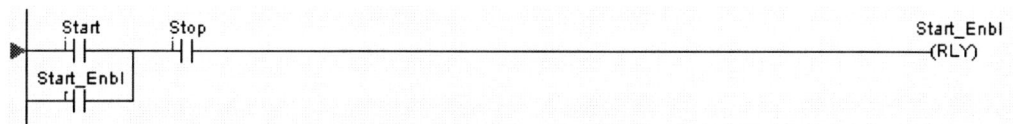

Fig 3.19

Transitions

Fig 3.20

States

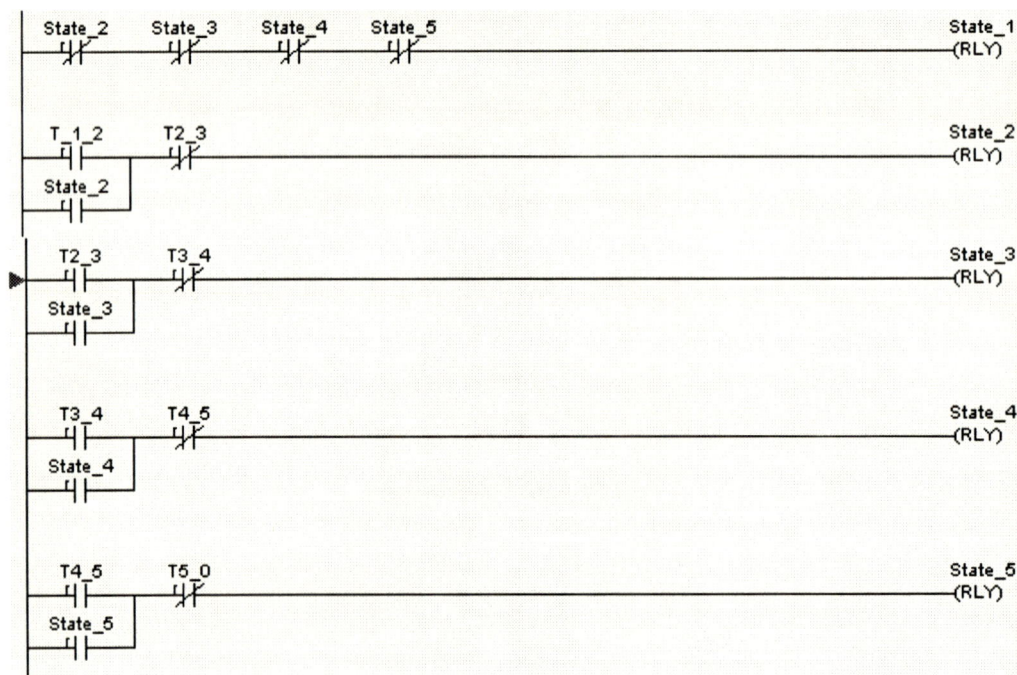

Fig 3.21

Outputs

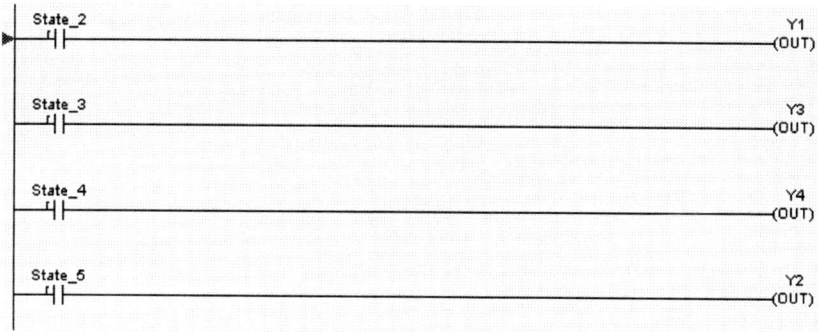

Fig 3.22

3.6 Modifying a SFC Program

Lets imagine now that the the directional control valve (DCV) controlling cylinder A is to be replaced and the only available replacement is a solenoid operated, spring return 5/2 DCV.

The modified pneumatic diagram is shown in *Fig 3.23,* and the PLC wiring diagram in *Fig 3.24*

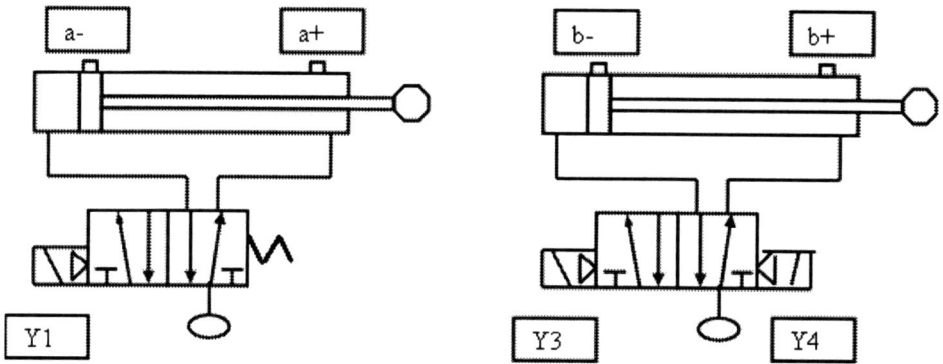

Fig 3.23

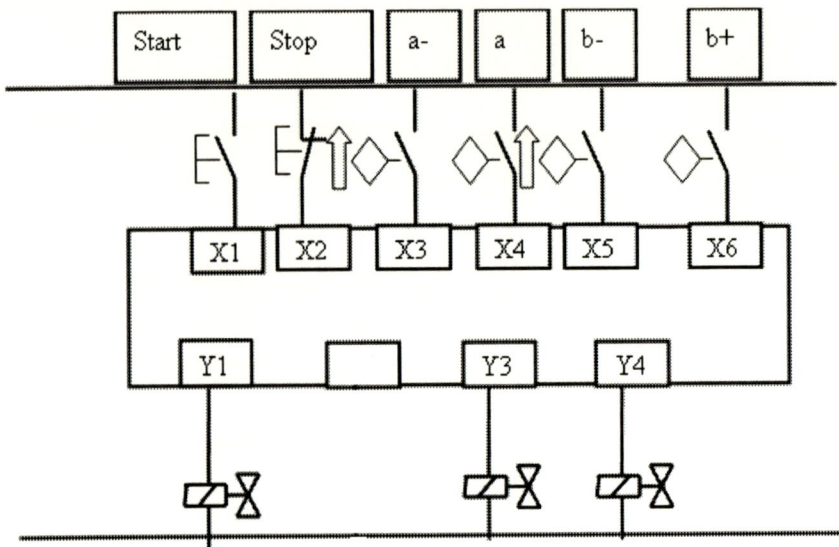

Fig 3.24

The cylinder sequence remains unchanged so cylinder A still extends during State 2 and must stay extended during states 3 and 4.

The only part of the ladder logic that needs to be modified is the output section. This is shown in *Fig 3.25*.

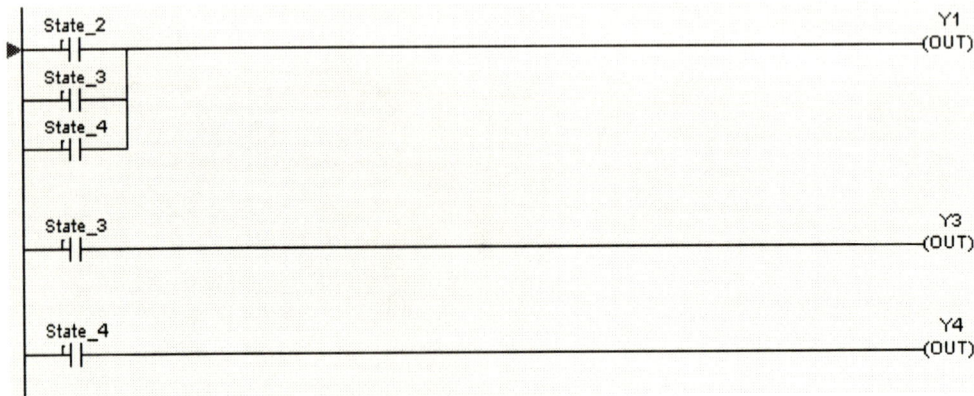

Fig 3.25.

3.7 Use of Timers and Counters in SFC Programming

In the two examples that we have seen (*3.1* and *3.2*) PLC inputs have been used to enable transitions from one state to the next. Sometimes timers or counters are used instead of inputs. For example, a container of liquid is stirred for 20 seconds or a shaft rotates 5 times to index a conveyor.

We'll start with an example that includes a timer. We can use the same hardware as *Example 3.1*

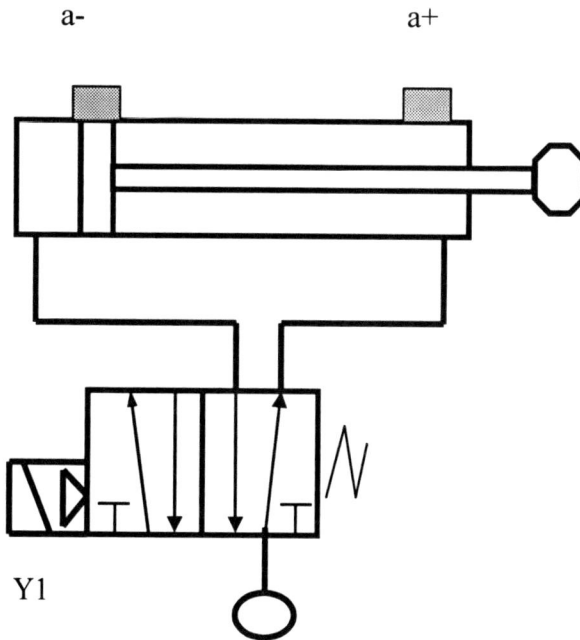

Fig 3.26

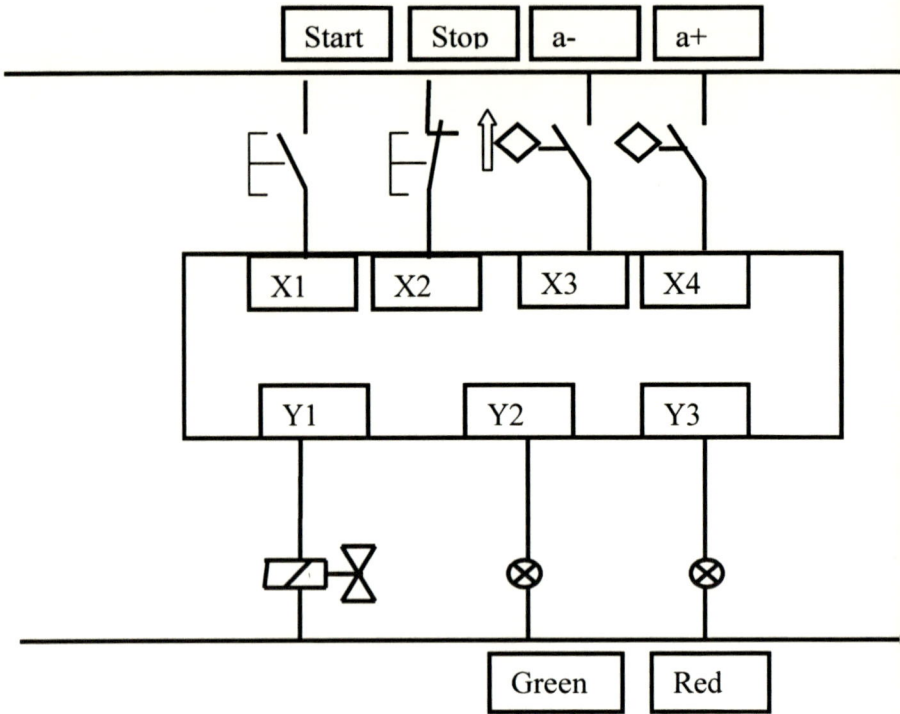

Fig 3.27

Example 3.3 The cylinder shown in *Fig 3.26* is to extend, remain extended for 3 seconds and then return. While the cylinder is extended, the red light Y3 is on and while it is retracted the green light Y2 is on. The SF chart is shown in *Fig 3.27*.

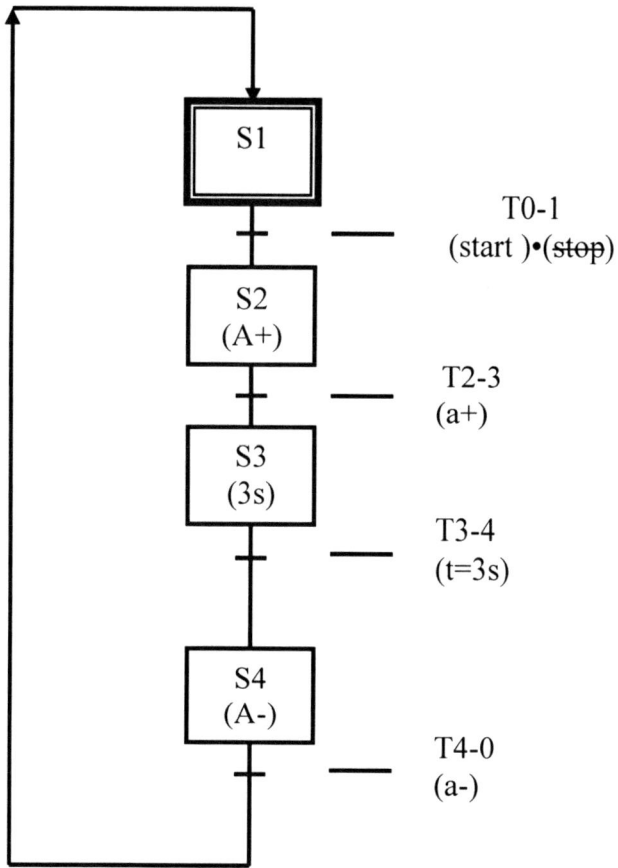

Fig 3.27

The assignment list is shown in *Table 3.4*

S1	R1
S2	R2
S3	R3
S4	R4
T1-2	R5
T2-3	R6
T3-4	R7
T4-0	R8

Table 3.4

Initial Conditions:

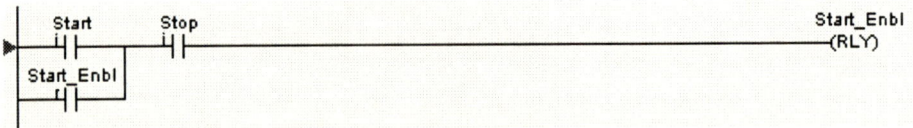

Fig 3.28

Transitions:

Fig 3.29

64

States:

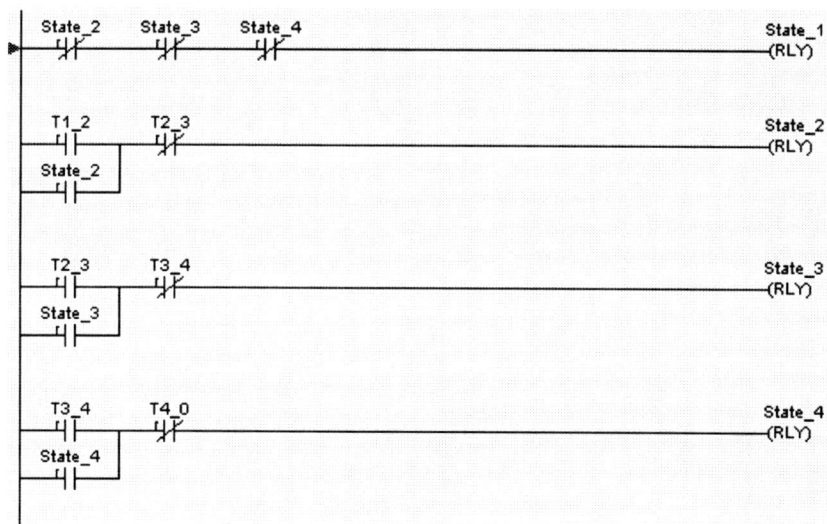

Fig 3.30

Outputs:

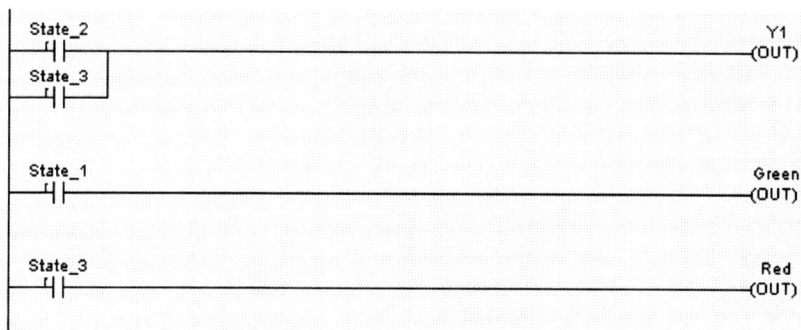

Fig 3.31

Timer:

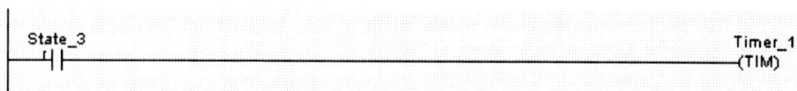

Fig 3.32

65

So when State 3 becomes active the timer starts (*Fig 3.32*). When the delay has finished a normally open contact of the timer is used to activate Transition 3-4 (*Fig 3.29*).

Now we'll do a counter example.

Example 3.4

An indexing conveyor carries empty oil containers for filling and capping as shown in *Fig 3.33*. The conveyor is indexed one pitch by three revolutions of the cam on the drive shaft as shown in *Fig 3.34*.

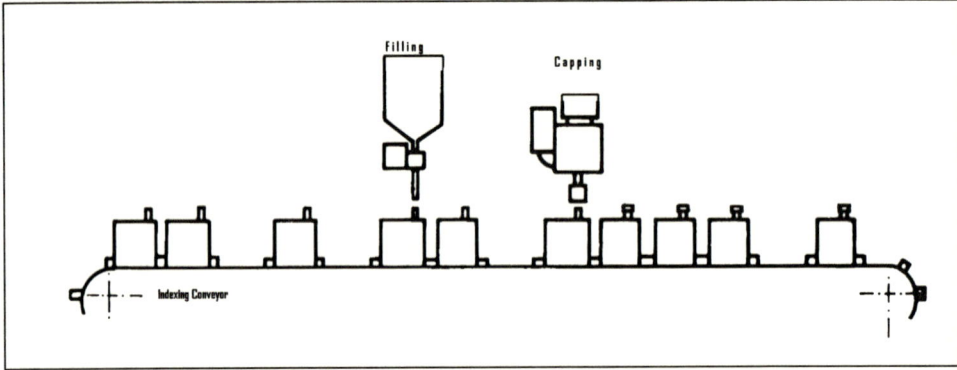

Filling   Capping

Indexing Conveyor

Fig 3.33

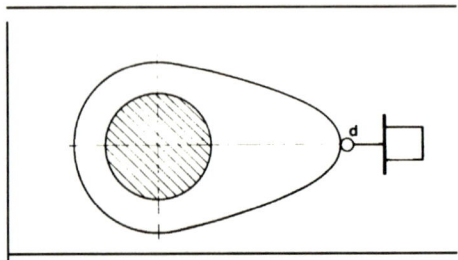

d

Fig 3.34

Write a PLC program that indexes the conveyor by one pitch when a start button is pressed.

The SF chart, PLC wiring diagram and assignment list are shown below.

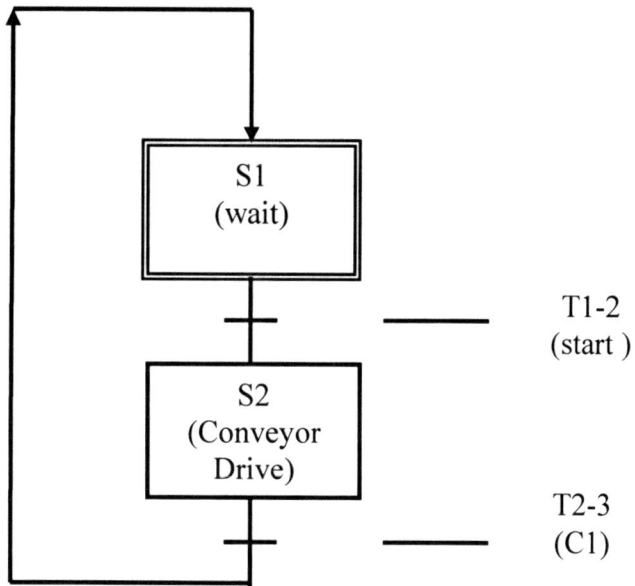

Fig 3.35 SF Chart

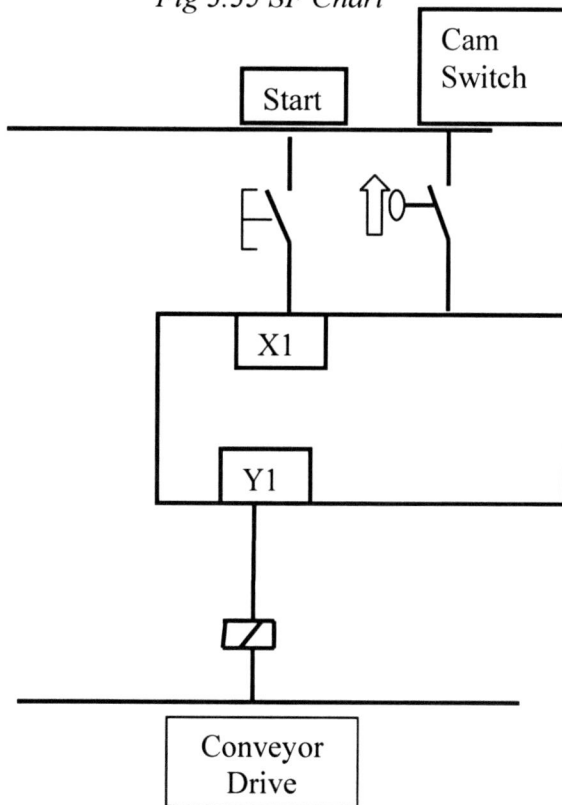

Fig 3.36 PLC Wiring Diagram

67

S1	R1
S2	R2
T1-2	R5
T4-0	R8

Table 3.5 Assignment List

Transitions:

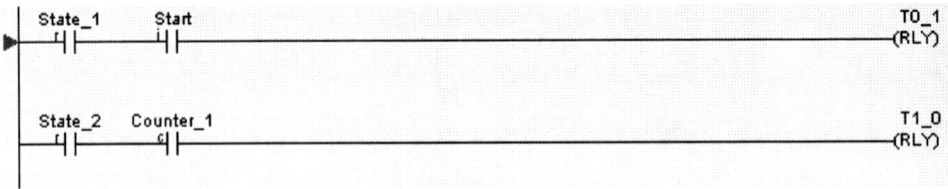

Fig 3.37

States:

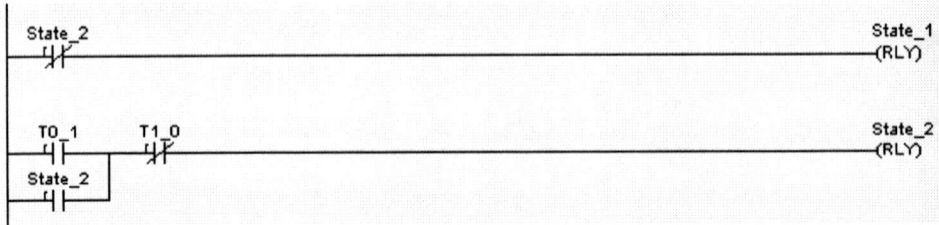

Fig 3.38

Outputs

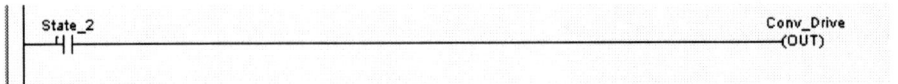

Fig 3.39

Counter

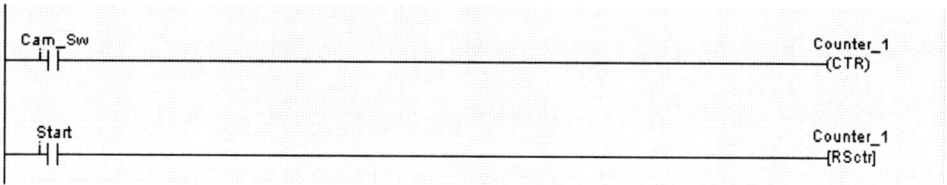

Fig 3.40

During State 2 the conveyor drive is energised (*Fig 3.39*) and the closing of the cam switch increments the counter. The counter is reset by the pressing of the *Start* button (*Fig 3.40*).

1.

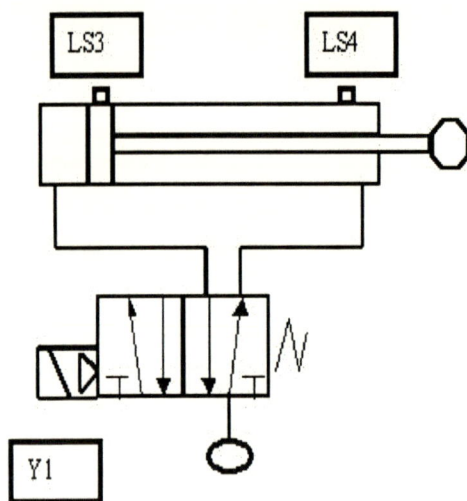

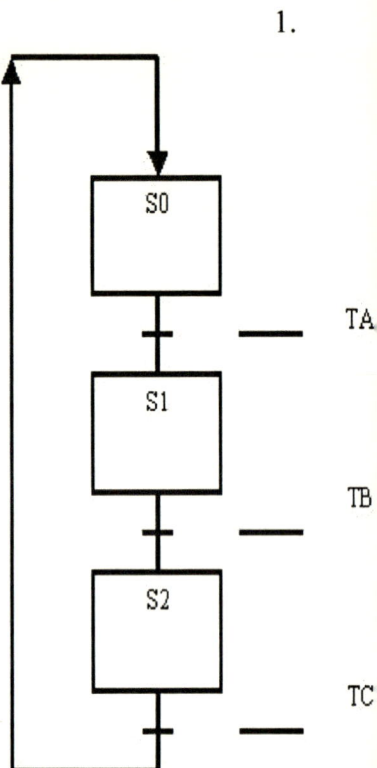

The dcv and cylinder in *Fig 3.41* are controlled by a plc with the program shown in *Fig 3.43*. If the 5/2 solenoid/spring dcv was replaced by a 5/2 double solenoid dcv and the second solenoid, Y2, connected to plc output Y2, then which statement below correctly describes during which states the outputs Y1 and Y2 would be energised?

Answer:

        a. Y1 in states 1 and 3, Y2 in state 2

        b. Y1 in states 1 and 3, Y2 in states 0 and 2

        c. Y1 in state 0, Y2 in state 3

        d. Y1 in state 1, Y2 in state 2

2.

```
   |R1      R2      R3                                              RO        |
 0 |—|/|——|/|——|/|——————————————————————————————————————————————( )———|
   |State 1 State 2 State 3                                      State 0      |
```

Select the only <u>correct</u> statement regarding the diagram *Fig 3.43*.

Answer:

    ◌   a. A latch for state 0 is missing in the diagram.

    ◌   b. State 0 is energised by transitions

    ◌   c. State 0 will be energised when the plc is first powered up

    ◌   d. State 0 is switched on when all the other states are on

3. Select the only <u>incorrect</u> statement below regarding the diagram *Fig 3.44*.

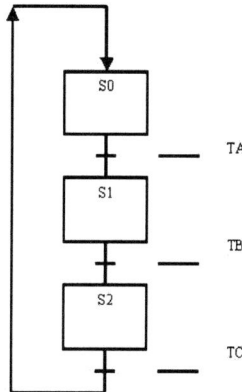

Fig 3.44

Answer: ◌   a. The states and transitions are represented in ladder logic by internal relays.

    ◌   b. States are latched on with the exception of state 0

    ◌   c. A transition is enabled by an input or timer contact.

    ◌   d. TA, TB and TC are plc inputs

4.

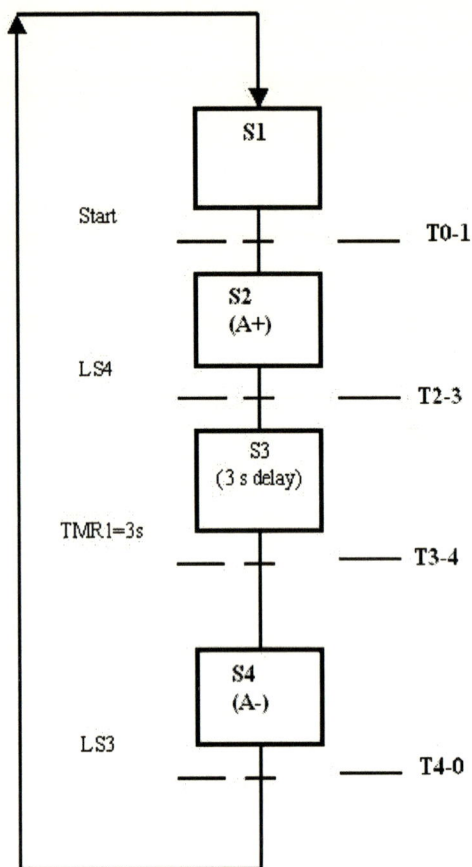

Fig 3.45

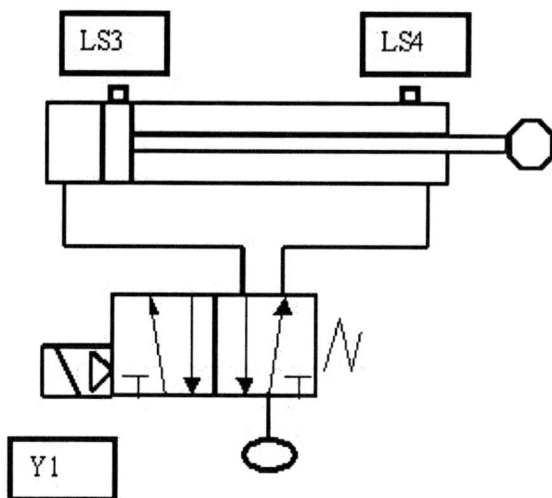

Fig 3.46

Select the only <u>correct</u> statement below regarding *Fig 3.45* which is used to control the pneumatic cylinder A in *Fig 3.46*.

Answer:

- ○ a. Input LS4 enables Transition 4-0
- ○ b. No output is switched on during state 3
- ○ c. The internal relay S3 is used to energise the timer.
- ○ d. No input is required for transition 0-1.

5.

Fig 3.47

Select the only <u>incorrect</u> statement below regarding *Fig 3.47*.

Answer: ○   a. State 1 is switched on by transition T1-2

           ○   b. State 1 is switched on by Transition T0-1

           ○   c. State 1 is latched because the transitions are only energised for brief periods of time.

           ○   d. State 1 is switched off by transition T1-2

6.

Fig 3.48

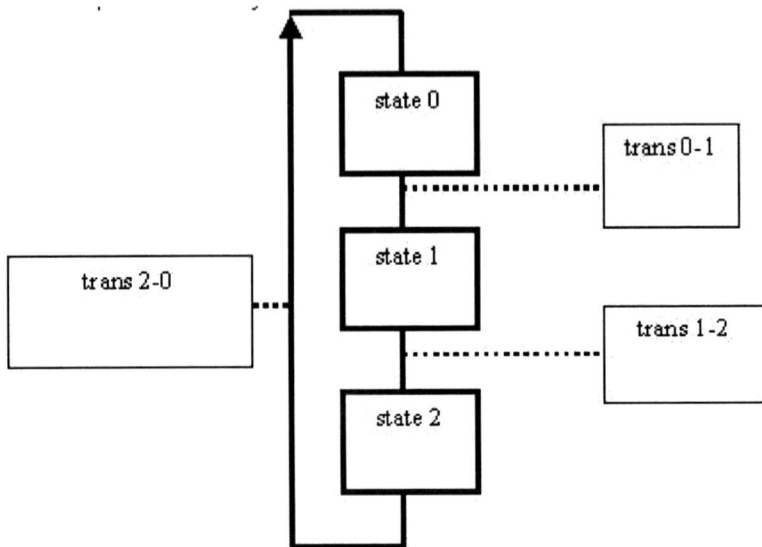

Fig 3.49

74

Which option below correctly identifies the fault with the ladder conditions shown in *Fig 3.48* for state 2 of *Fig 3.49*?

Answer:　　　　○　　　a. The state should be latched on.

　　　　　　　　○　　　b. State 2 is the default state.

　　　　　　　　○　　　c. The required output should be included.

　　　　　　　　○　　　d. An input or timer contact should be included in the conditions

7.

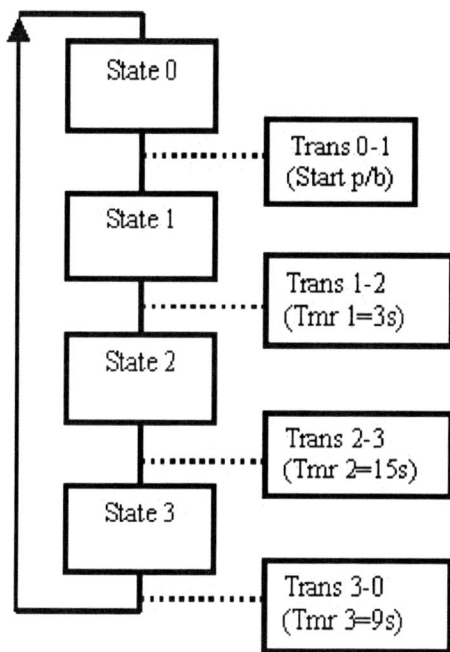

State 0

Trans 0-1
(Start p/b)

State 1

Trans 1-2
(Tmr 1=3s)

State 2

Trans 2-3
(Tmr 2=15s)

State 3

Trans 3-0
(Tmr 3=9s)

Select the only <u>incorrect</u> statement below regarding *Fig 3.50*

Answer: ○   a. State 2 is active for 15 seconds.

○   b. When the plc is first powered up state 0 becomes active.

○   c. It is not possible to mix transitions enabled by inputs and transitions enabled by timers in the same program.

○   d. When trans 3-0 becomes active state 3 switches off.

8.

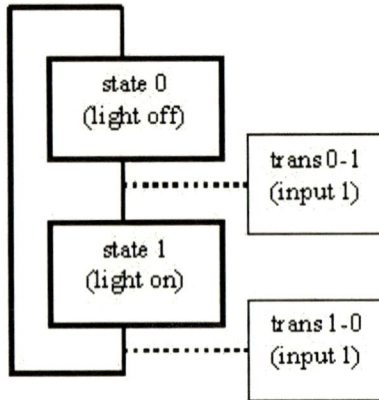

Fig 3.51

Select the only correct statement below regarding Fig 3.51

Answer: ○   a. Input 1 only switches the light off.

○   b. Input 1 only switches the light on.

○   c. There can be no default state in such a sequence.

○   d. Input 1 alternately switches the light on and off.

76

9.

It is necessary to automate the drilling of 2 holes in the clamp body shown in *Fig 3.52*. Both holes are at an angle of 45° to the vertical .

For the drilling operation, the clamp is mounted on the rotation jig shown in *Fig 3.53*

Two views of the drill assembly are shown in *Fig 3.54*. Each double acting pneumatic cylinder is operated by a double solenoid 5/2 directional control valve.
Limit switches are positioned as shown to detect the advanced and retracted cylinder positions.

When the start button is pressed the sequence is as follows

1. The first hole is drilled.
2. The jig is unclamped by cylinder I.
3. The jig is rotated through 90° by the rack on cylinder R.
4. The jig is clamped (cylinder I).
5. The second hole is drilled.
6. The jig is unclamped (cylinder I)
7. The jig is rotated back through 90° (cylinder R).
8. The jig is clamped (cylinder I).

The drill motor runs continuously throughout the sequence.

(a) Draw the appropriate state/transition diagram.

(b) Draw a suitable plc wiring diagram assigning I/Os as necessary.

(c) Draw the lines of ladder logic to operate the system as designed.

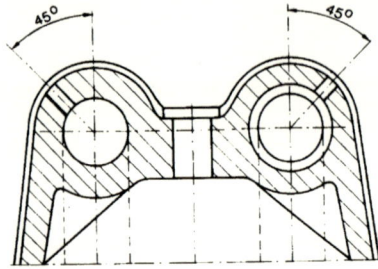

Fig 3.52. Clamp Body

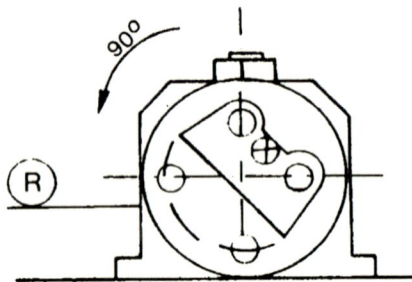

Fig 3.53 Rotation Jig

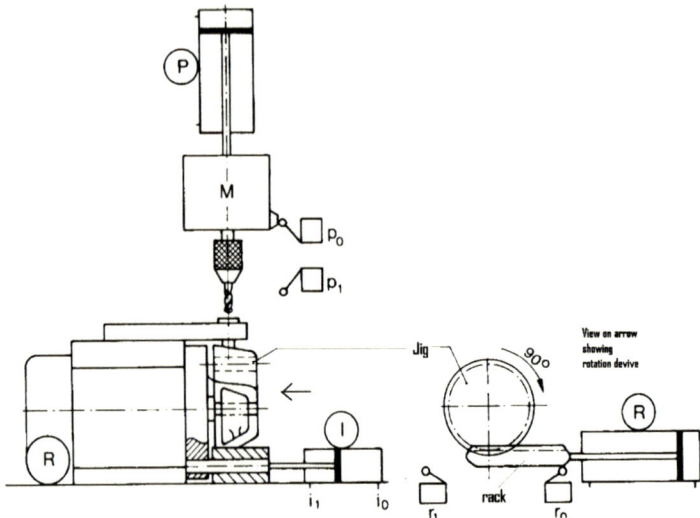

Fig 3.54 Drill Assembly

78

10.

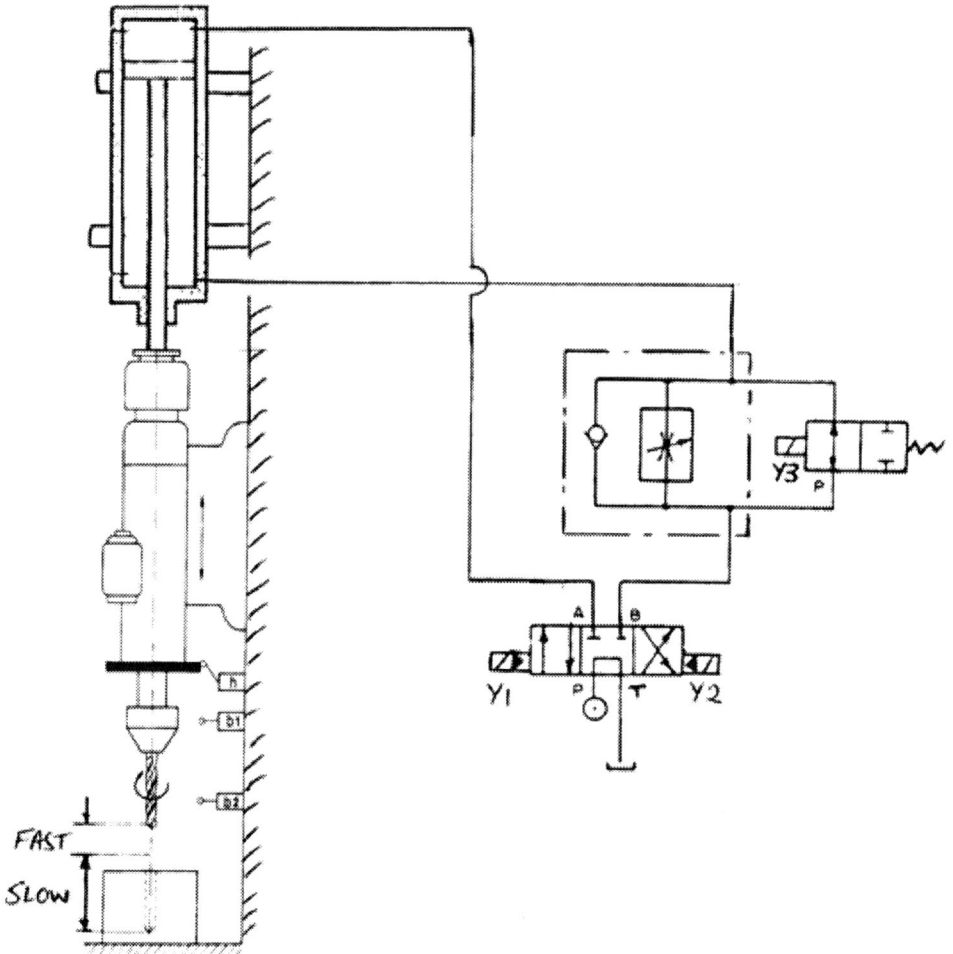

The diagram *Fig 3.55* shows a quick approach circuit for a drill. When a *start* push button is pressed the drill motor starts and the drill assembly quickly approaches the work piece. When the limit switch b1 is reached the drill continues slowly until the end-of-stroke limit switch b2 is closed. The assembly then returns to the top position and the drill motor stops. The hydraulic pump may be assumed to run continuously.

(a) Draw a plc wiring diagram for the system.
(b) Draw a suitable state/ transition diagram.
(c) Draw the ladder logic to operate the system as designed.

Chapter 4
Selective Branching

4.1 Introduction

It is often necessary for a production sequence to take one route or another based on a decision factor. E.g. if a product is good the process goes on to the next operation, if it is bad the product is rejected. SF chart programming copes easily with this sort of situation.

Example 4.1

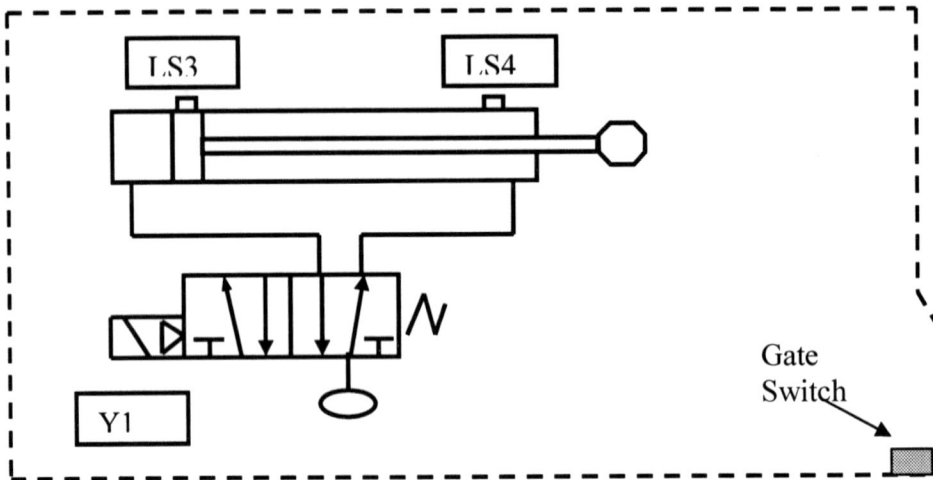

Fig 4.1

When a Start pushbutton is pressed and the gate is closed the cylinder in *Fig 4.1* is to extend for 3 seconds and then return. If the start button is pressed but the gate is not shut, an alarm sounds until the gate is closed. A *cycle on* light is lit during the time that the cylinder is not retracted.

The PLC wiring diagram is shown in *Fig 4.2* and the SF chart in *Fig 4.3*

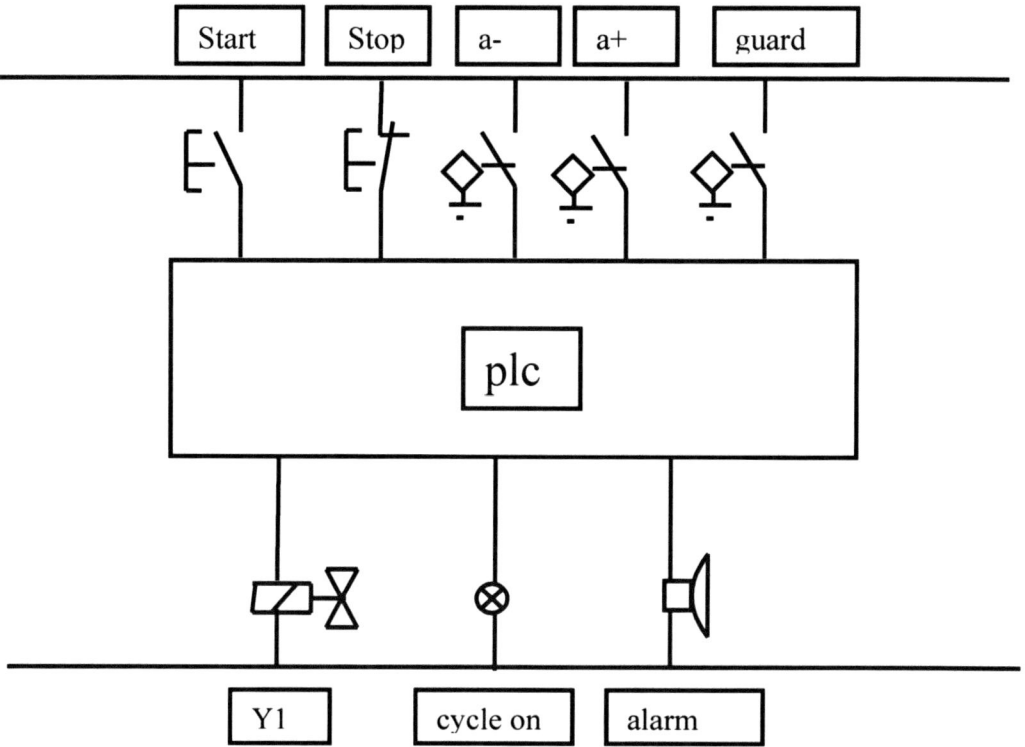

Fig 4.2

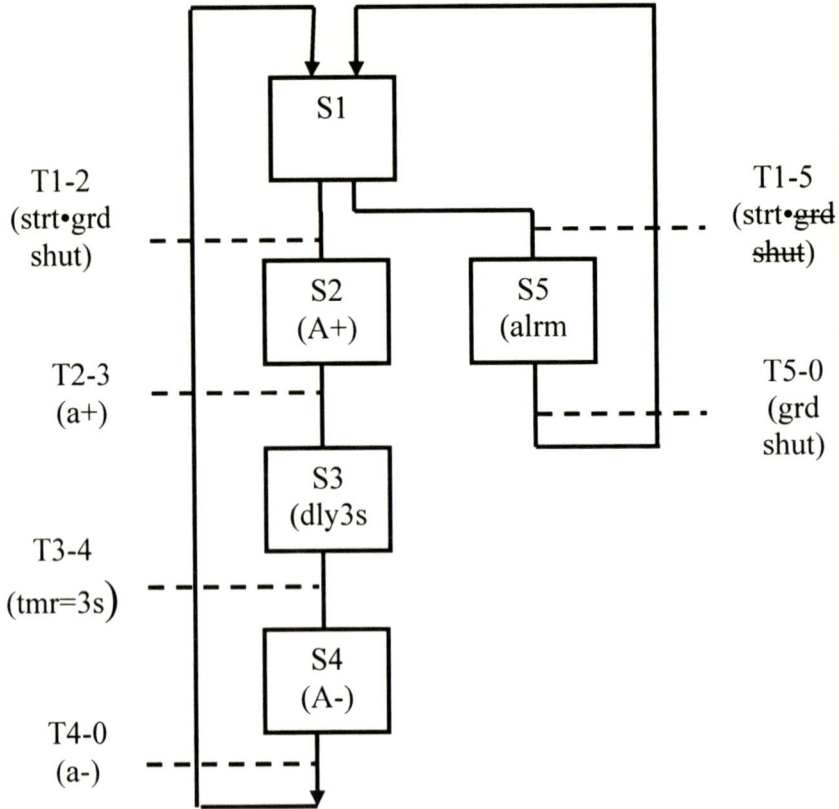

Fig 4.3

If the guard switch is closed and the *start* button is pressed then the transition 1-2 is activated and the cylinder cycle starts. If, however the guard is open and the *start* button is pressed then transition 1-5 becomes active and the alarm sounds until the guard is closed.

The assignment list is shown in *Table 4.1*

S1	R1
S2	R2
S3	R3
S4	R4
S5	R5
T1-2	R6
T2-3	R7
T3-4	R8
T4-0	R9
T0-5	R10
T5-0	R11

Table 4.1

And so we can write the ladder logic.

Initial Conditions:

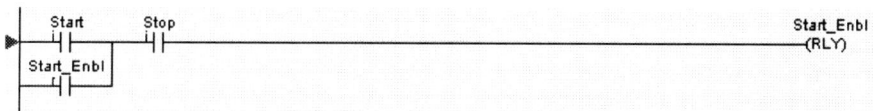

Fig 4.4

Transitions:

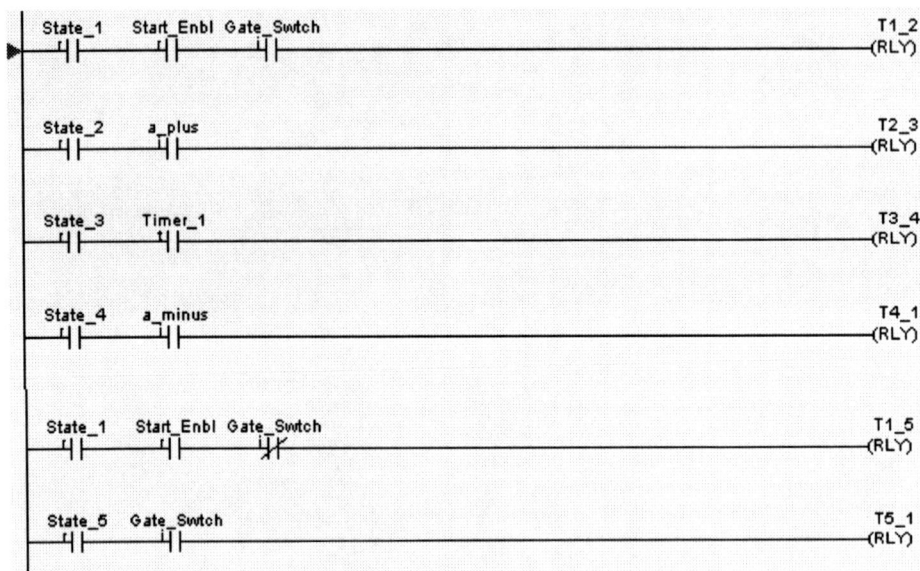

Fig 4.5

States:

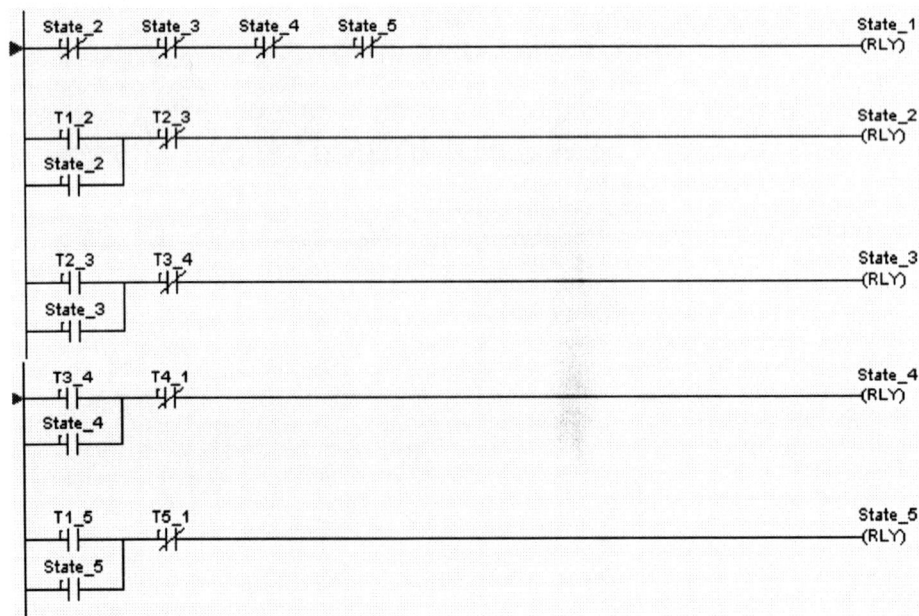

Fig 4.6

84

Outputs:

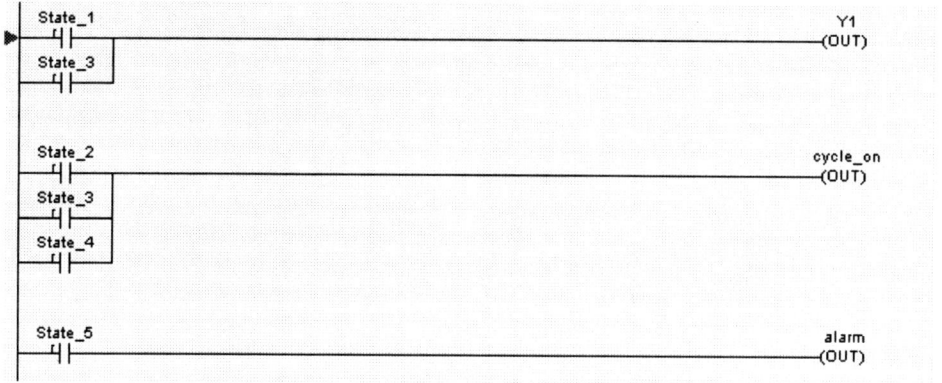

Fig 4.7

Timer:

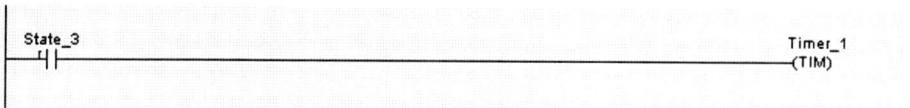

Fig 4.8

It is important to remember that the decision criteria whether to choose one branch or another must be mutually exclusive. In this case it is the logic state of the gusrd switch. This is a suitable criterion because the switch cannot be simultaneously open and closed.

85

Example 4.2

Here is another example that uses the selective branching technique.

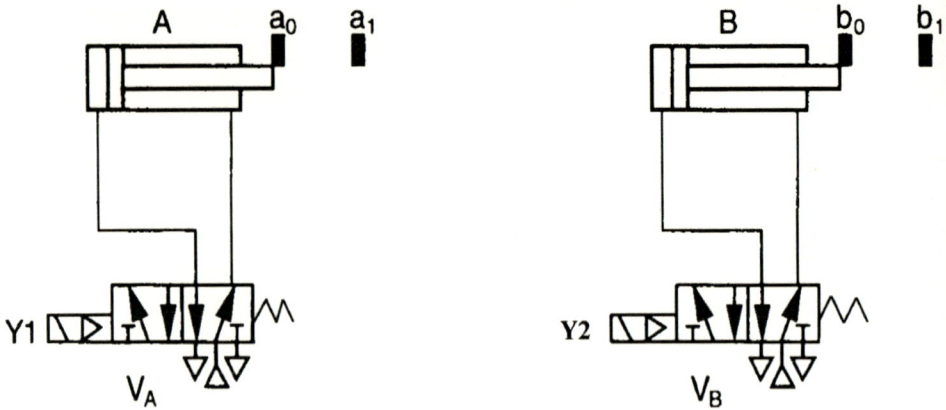

Fig 4.9

The cylinders above are to execute the sequence A+ B+ B- A- continuously until a stop button is pressed when both cylinders will stop at the end of the sequence.

If, however a self-latching switch is turned on, the cycle becomes A+, 5s delay, A-, executed continuously as before.

(a) Draw a plc wiring diagram for the system.

(b) Write out an assignment list for the system

(c) Draw a sequential function chart for the system.

(d) Draw the ladder logic to operate the system as designed.

Solution

(a)

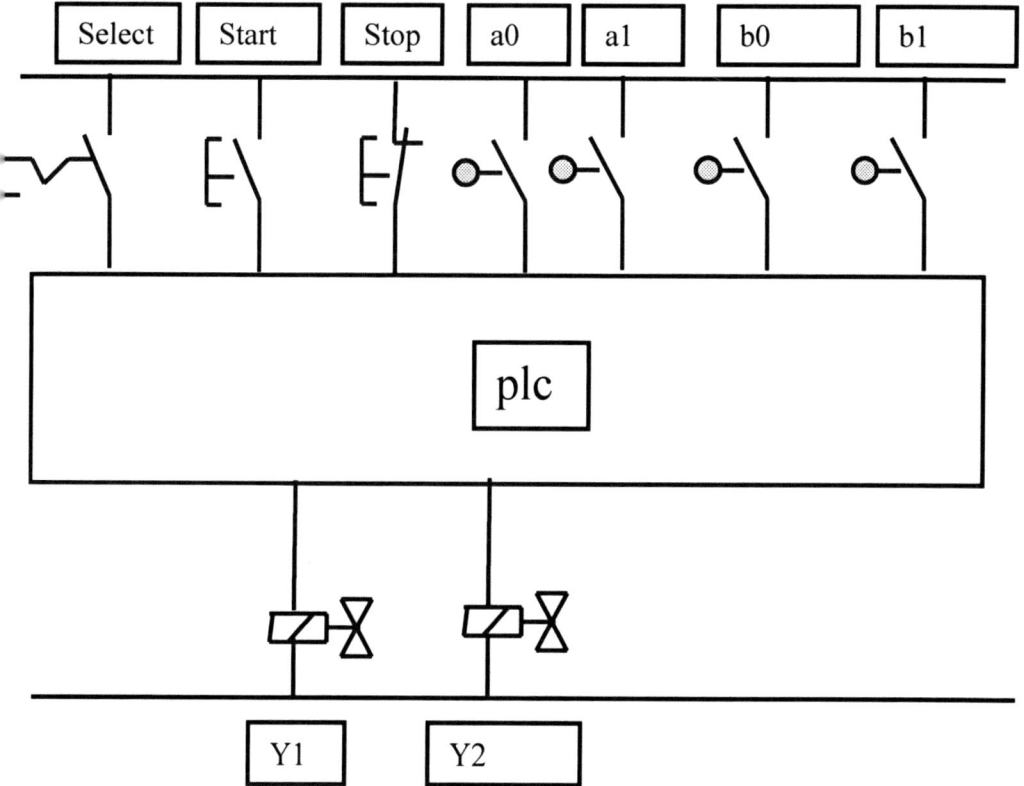

Fig 4.10

(b)

S1	R1	T1-2	R9
S2	R2	T2-3	R10
S3	R3	T3-4	R11
S4	R4	T4-5	R12
S5	R5	T5-1	R13
S6	R6	T1-6	R14
S7	R7	T6-7	R15
S8	R8	T7-8	R16
		T8-1	R17

Table 4.2

(c)

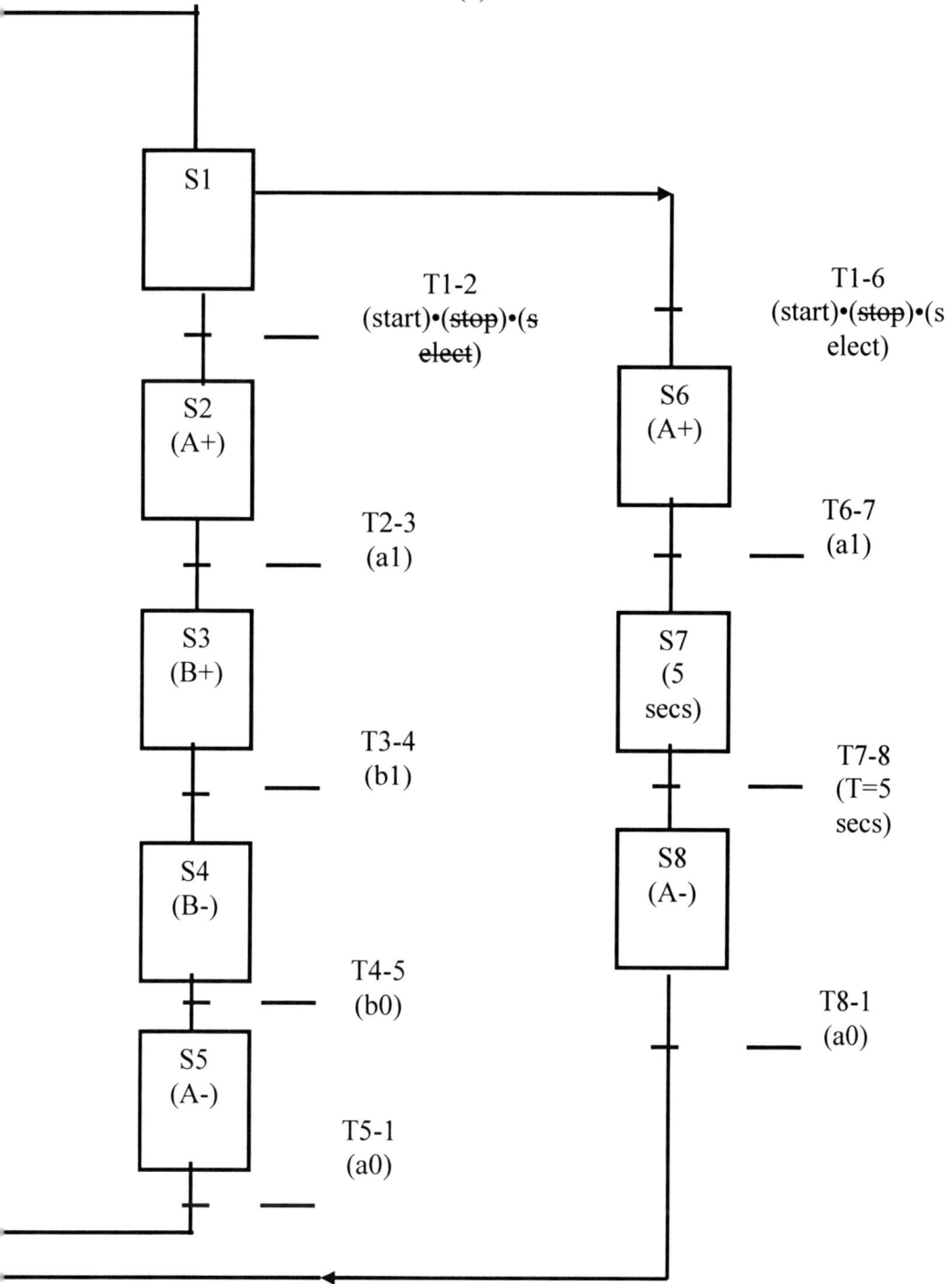

Fig 4.11

(d)

Initial Conditions:

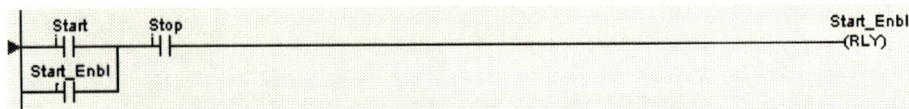

Fig 4.12

Transitions:

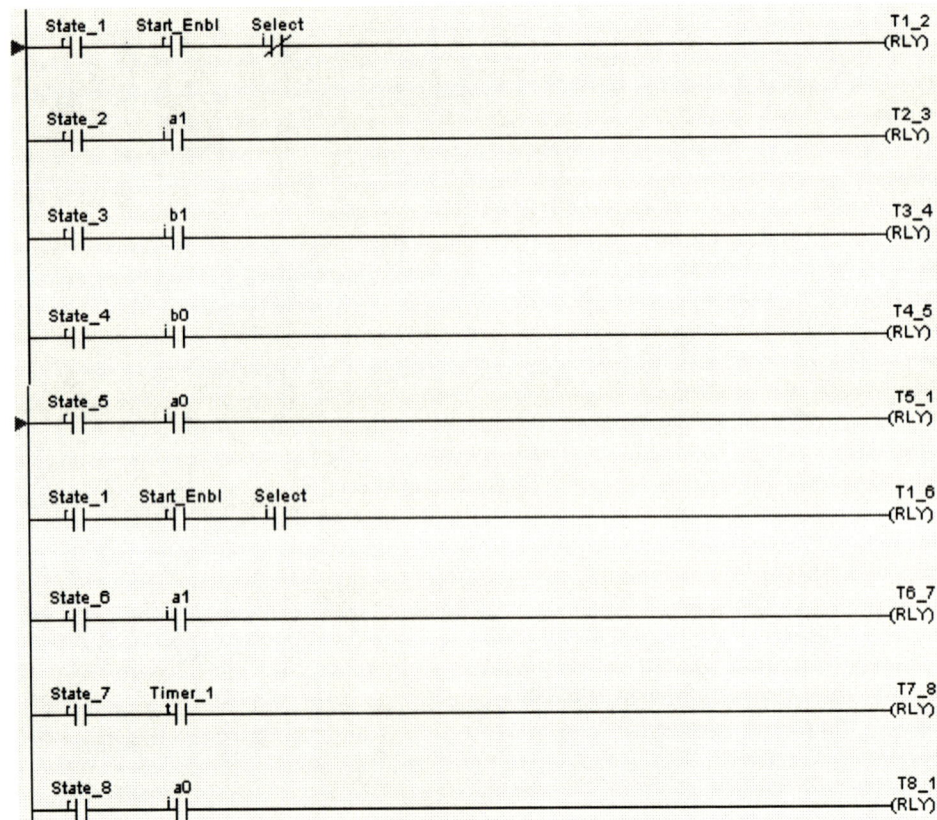

Fig 4.13

States:

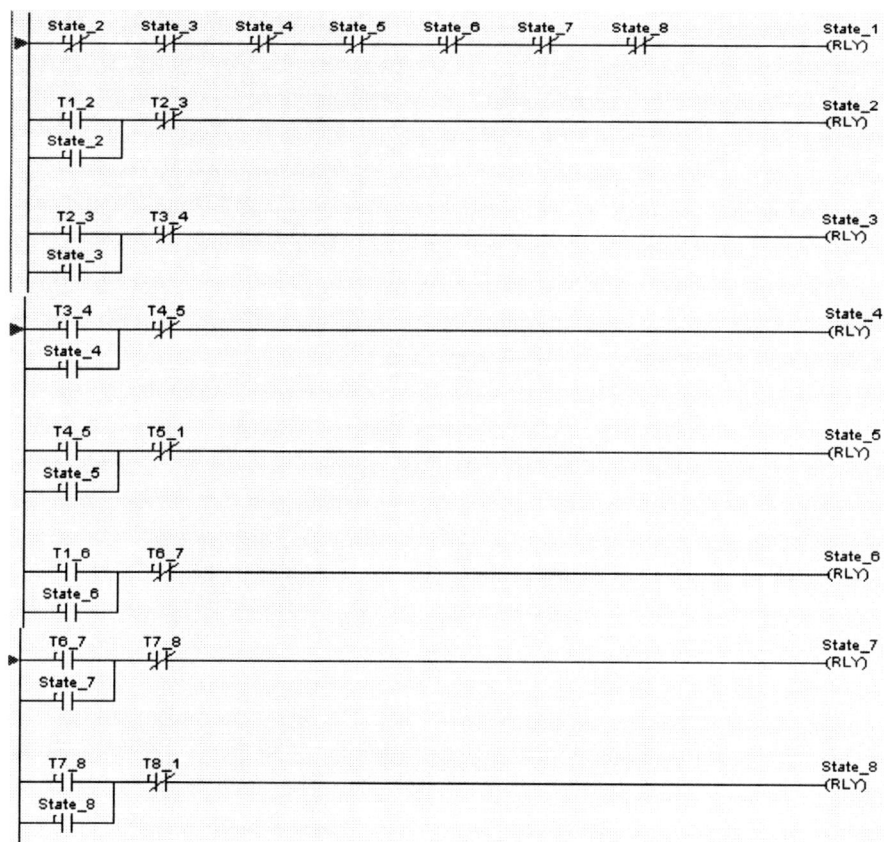

Fig 4.14

Outputs:

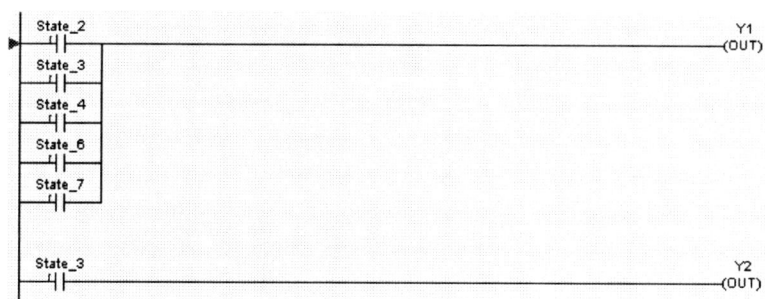

Fig 4.15

Timer:

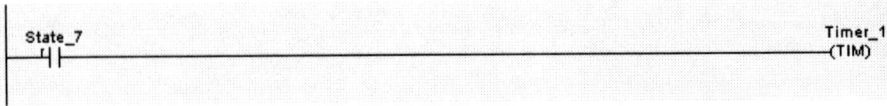

State_7   Timer_1
                                                                                           (TIM)

Fig 4.16

Example 4.3

Here is a slightly more complicated example of a process that uses selective branching. The solution is straightforward provided we stick to the method for SF chart programming.

A process liquid is prepared in a tank as shown in *Fig 4.9*. The tank is filled with liquid from pump A until the high level switch is reached, then a small amount of liquid is added by running pump B for 10 seconds.. The mixture is stirred and heated to 60 ° C before being manually drawn off as required.

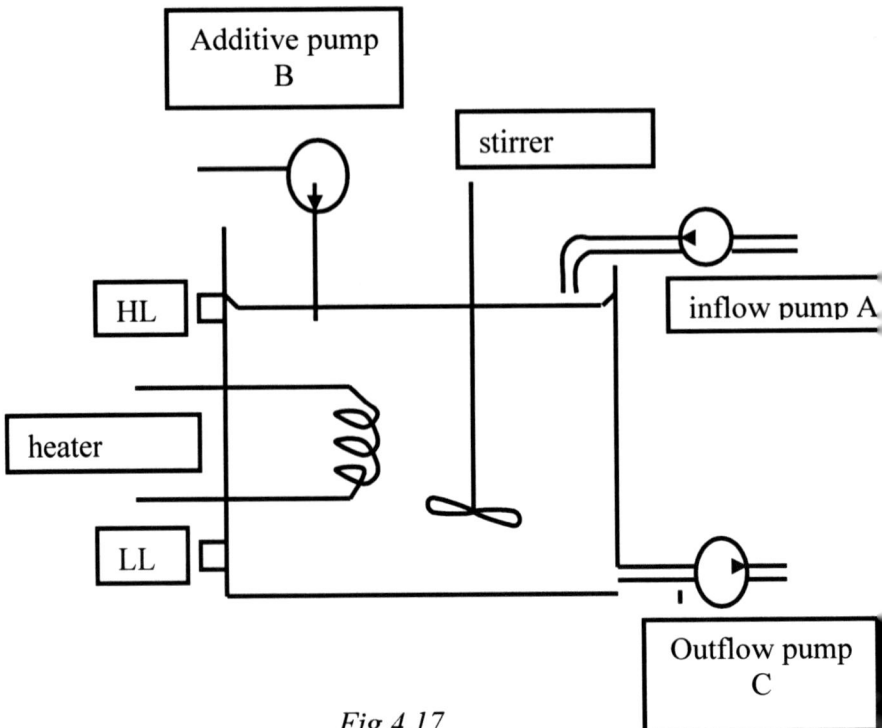

Fig 4.17

The SF chart is shown in *Fig 4.10*

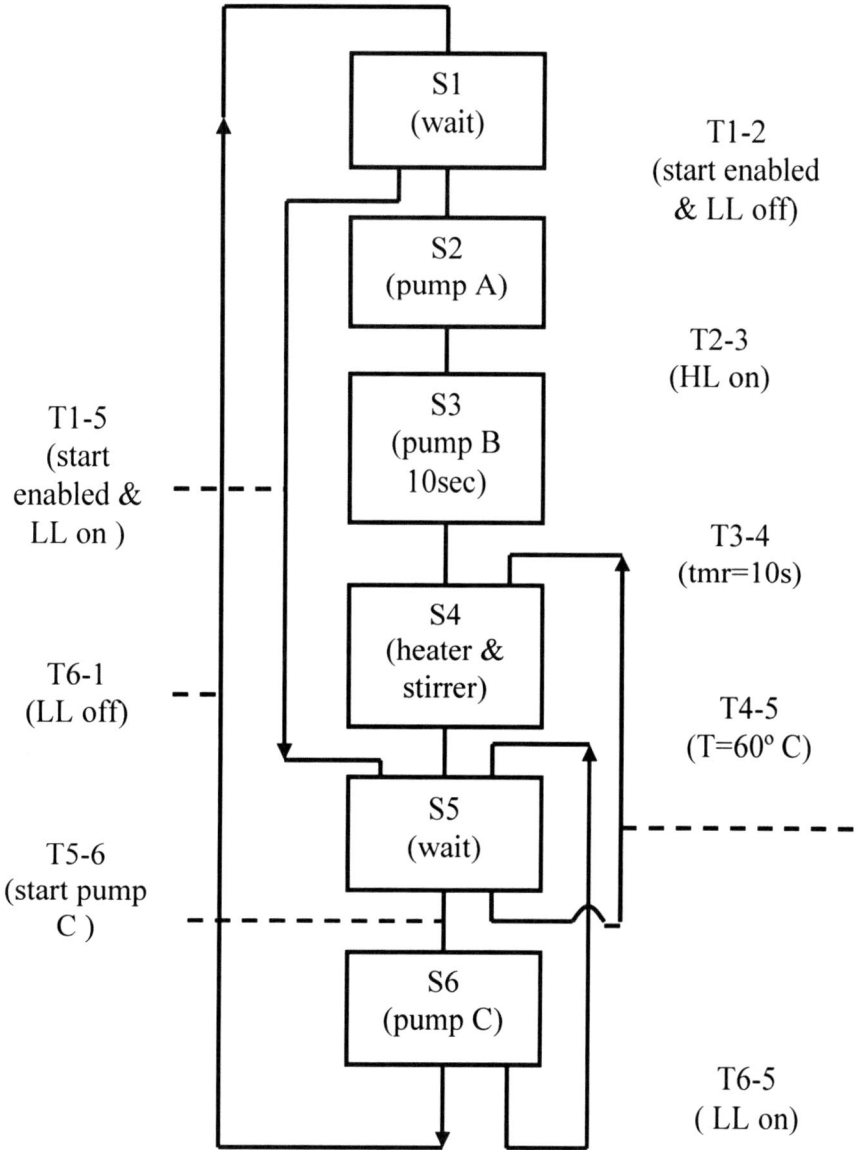

Fig 4.18

We will look at the various branches of the program.

We will first consider the case that the start button has been pressed and the tank is completely empty. The low-level (LL) switch being open, T1-2 is enabled and State 2 becomes active. During this state Pump A pumps liquid into the tank.

When the liquid level reaches the high-level (HL) switch, State 3 becomes active and Pump B runs for 10 seconds.
When the 10 seconds are up the contents of the tank are ready to be mixed and heated; this happens during State 4.

When the temperature of 60 ° C is reached, the system enters State 5, during which no output is energised.

When Pump C is switched on (manually) State 6 becomes active. If the tank is completely drained by pump C then State 1 switches on for the cycle to restart. If however pump C is switched off while there is still liquid left then State 5 becomes active again.

Another scenario occurs if, during State 5 the temperature of the mixture falls to 50 ° C. State 4 then becomes active and the mixture is heated and stirred until the temperature once again reaches 60 ° C.

The only remaining transition to consider is T1-5. This occurs is there is liquid in the tank at system start-up. When the start enable relay is energised state 5 becomes active followed by state 4 if the liquid temperature is less than 50 ° C or by state 6 if it is hot enough and pump C is switched on.

The PLC wiring diagram is shown in *Fig 4.11* and the assignment list in *Table 4.2*.

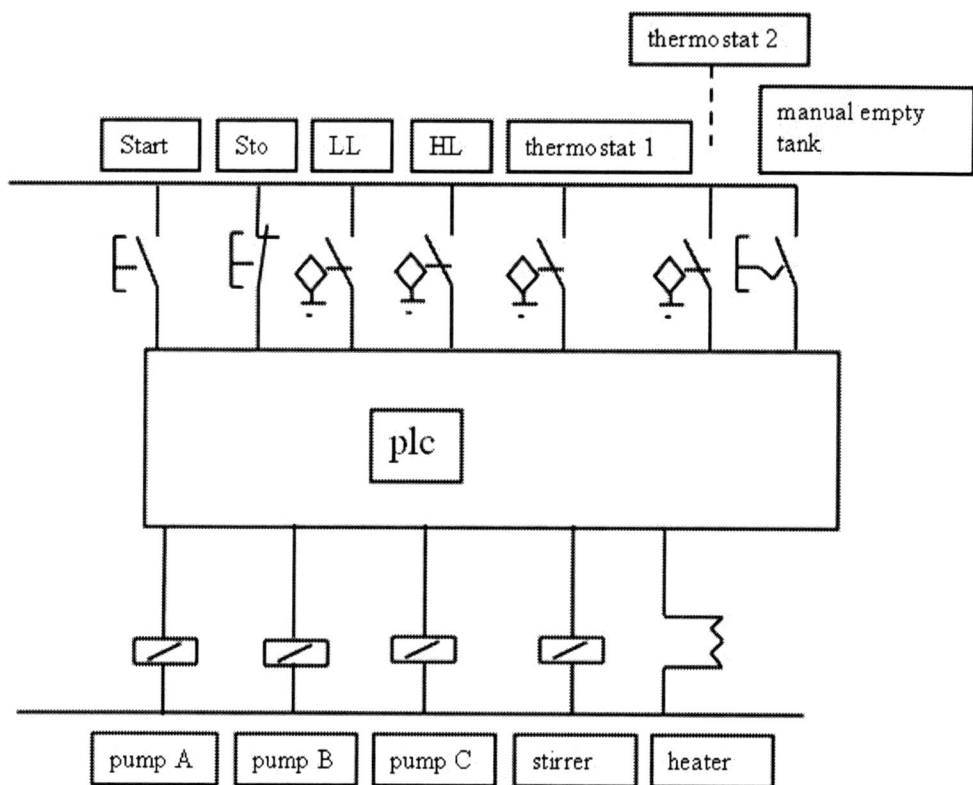

Fig 4.19

S1	R1
S2	R2
S3	R3
S4	R4
S5	R5
S6	R6
T1-2	R7
T2-3	R8
T3-4	R9
T4-5	R10
T5-6	R11
T6-1	R12
T1-5	R13
T5-4	R14
T6-5	R15

Table 4.3

The SF chart is a tool that the designer can use when discussing his proposals with the customer. The company safety officer might suggest the inclusion of panel lights or alarms, the process operator will have experience of practicalities which the designer may not have considered, the quality assurance manager may introduce measurement requirements etc.

When the design has been agreed, typing in the ladder logic is probably the easiest part of the job.

Initial Conditions:

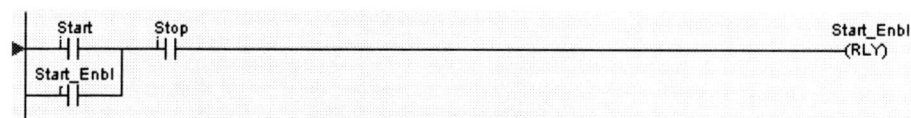

Fig 4.20

Transitions:

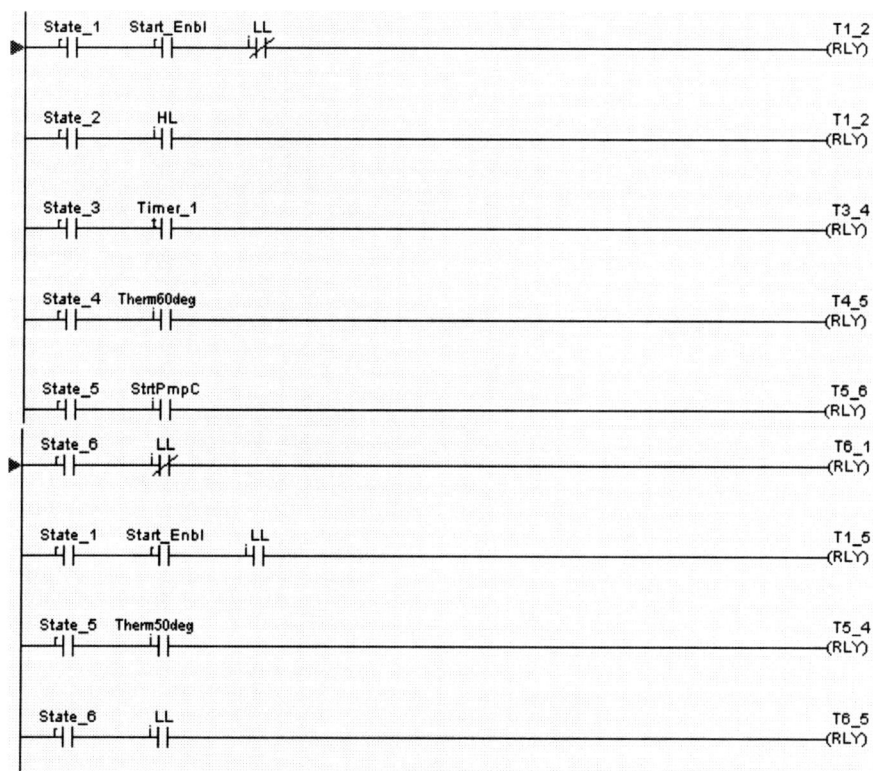

Fig 4.21

States:

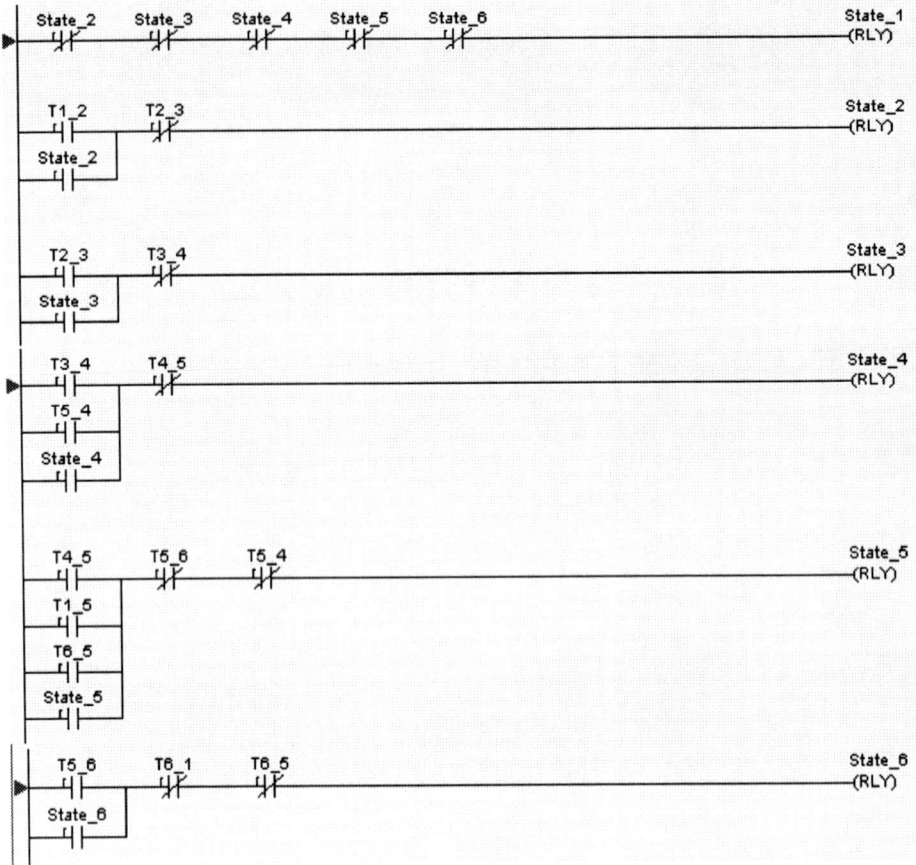

Fig 4.22

There are two routes into State 4 on the SF chart but only one route out of it. Stated another way, State 4 can be switched on by T3-4 or T5-4 and is switched off by T4-5. Similarly State 5 can be switched on by any of the three transitions T4-5, T1-5 or T6-5 and is switched off by T5-6 or T5-4. If you are in any doupt about this refer back to section 2.4 Outputs and Latches.

98

Outputs:

Fig 4.23

1.

Fig 4.25

Which option correctly identifies the fault with the ladder conditions shown in *Fig 4.25* for state 2 of the state/transition diagram shown in *Fig 4.26*?

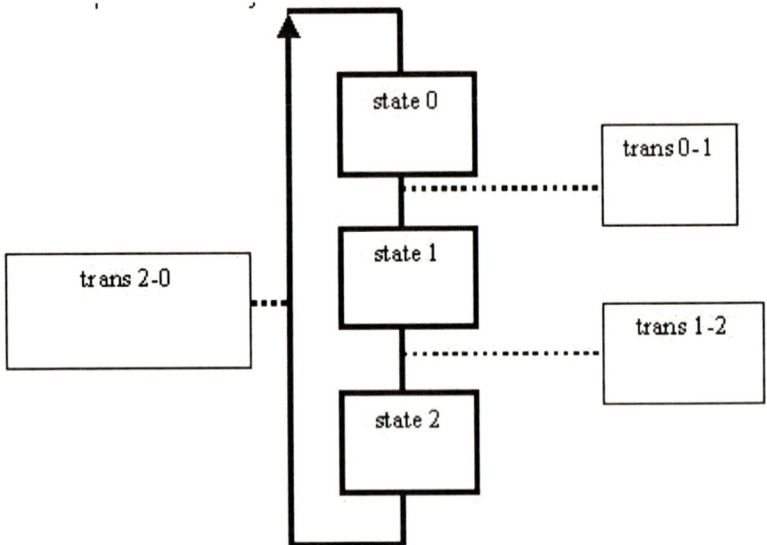

Fig 4.26

Answer: ○   a. The required output should be included.

              ○   b. State 2 is the default state.

              ○   c. The state should be latched on.

              ○   d. An input or timer contact should be included in the conditions.

2.

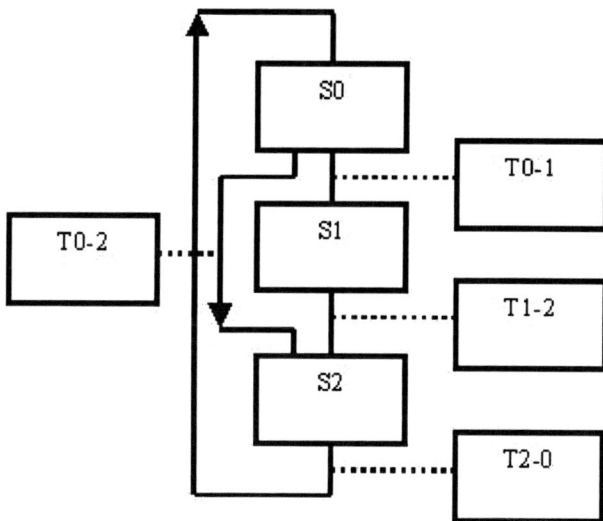

Fig 4.27

The ladder diagram in *Fig 4.28* is part of the sequential program in *Fig 4.27*.

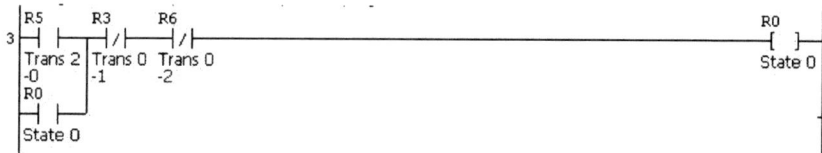

Fig 4.28

Select the one <u>correct</u> statement below regarding both.

Answer: ○   a. It is wrong to have a state contact on the left and over on the right as well.

○   b. State 0 is the default state so it shouldn't be specified in terms of transitions.

○   c. Trans 2-0 should be shown as a n/c contact not n/o.

○   d. Trans 0-1 and Trans 0-2 n/c contacts should be in parallel, not in series as shown.

101

3.

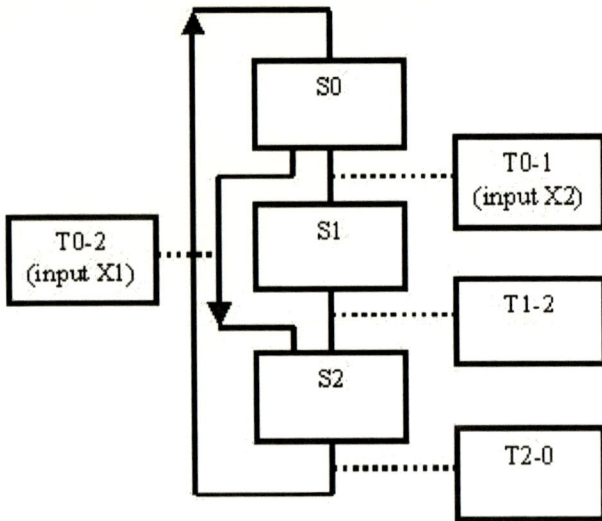

Fig 4.29

Select the only correct statement below about the state/transition diagram *Fig 4.29*.

Answer:
- ○ a. Trans 2-0 is unnecessary because state 0 is the default state.
- ○ b. It is not possible to skip a state in selective branching.
- ○ c. State 2 cannot work because there are two routes into it but only one route out.
- ○ d. The input conditions for trans 0-1 and trans 0-2 should be mutually exclusive

4.

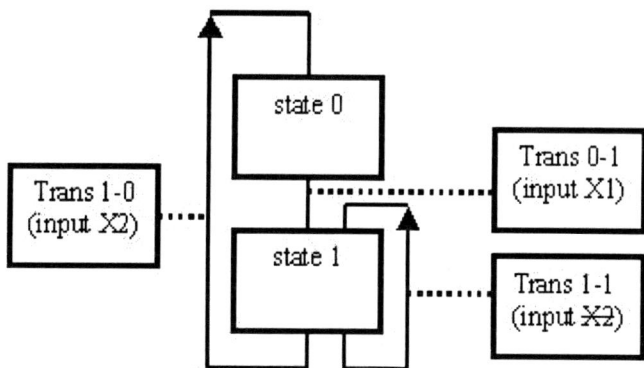

Fig 4.30

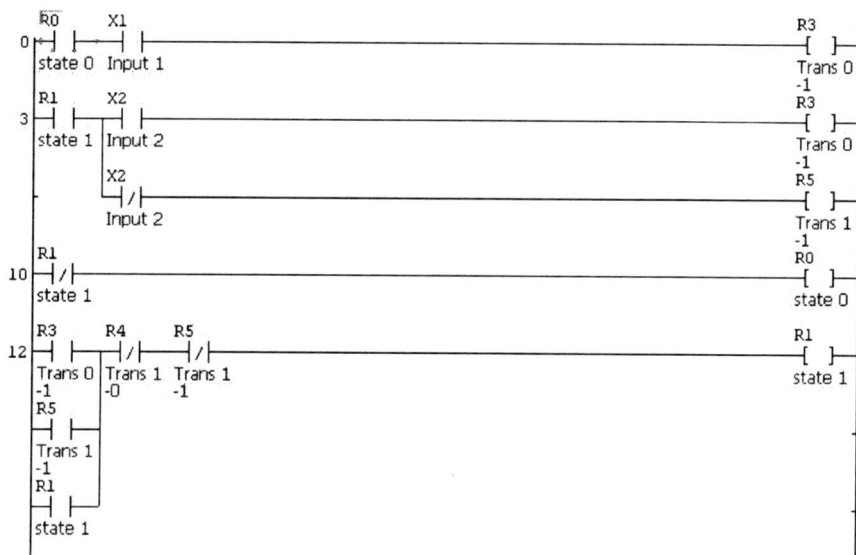

Fig 4.31

The ladder diagram *Fig 4.31* is part of the program for the state/transition diagram *Fig 4.30*. Select the only <u>correct</u> statement below regarding both.

Answer: ○ a. The input 1 contact should be shown n/c on the first line of ladder logic.

○ b. Trans 0 is the default transition.

○ c. State 1 cannot be a part-condition for two transitions as shown in the ladder diagram.

○ d. The ladder logic for for state 1 shows that you can't have a transition out of and back into the same state

5.

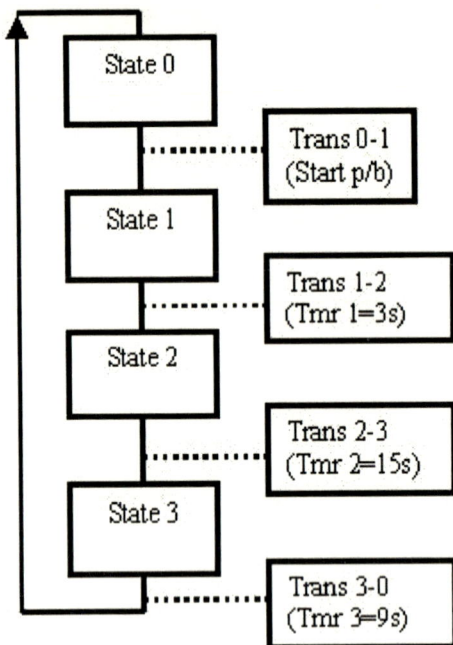

State 0

Trans 0-1
(Start p/b)

State 1

Trans 1-2
(Tmr 1=3s)

State 2

Trans 2-3
(Tmr 2=15s)

State 3

Trans 3-0
(Tmr 3=9s)

Fig 4.32

Select the only _incorrect_ statement below regarding the state/transition diagram *Fig 4.32*.

Answer: ○ a. When the plc is first powered up state 0 becomes active.

○ b. State 2 is active for 15 seconds.

○ c. When trans 3-0 becomes active state 3 switches off.

○ d. It is not possible to mix transitions enabled by inputs and transitions enabled by timers in the same program.

104

6.

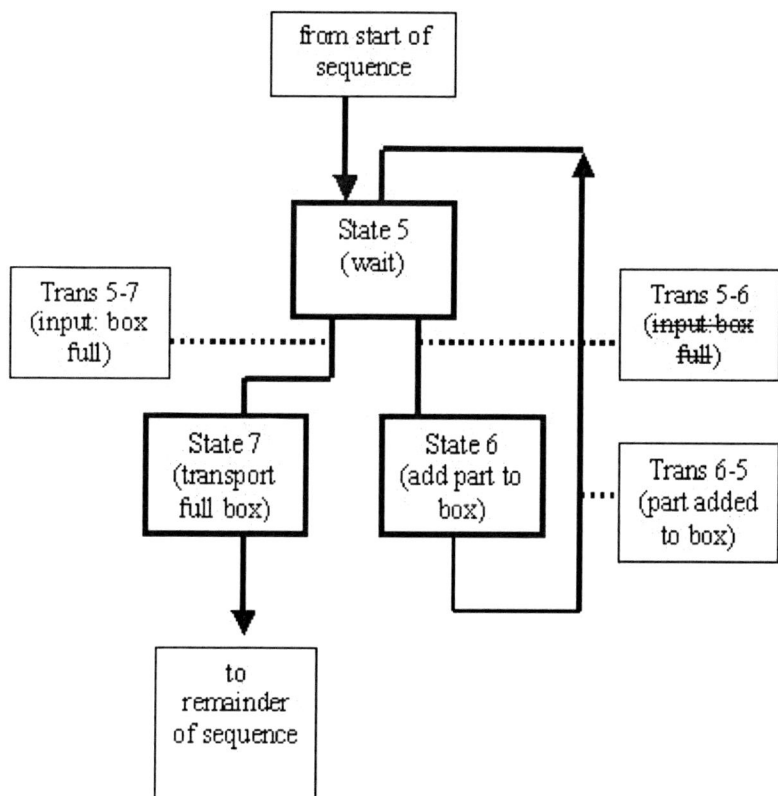

Fig 4.33

Select the only <u>incorrect</u> statement below regarding the diagram *Fig 4.33*

Answer: ○   a. When a box is full state 7 becomes active.

○   b. When a part has been added to a box trans 6-5 becomes active whether or not the box is full.

○   c. States 6 and 7 will never be simultaneously active.

○   d. Trans 5-7 and trans 5-6 are not mutually exclusive.

7.

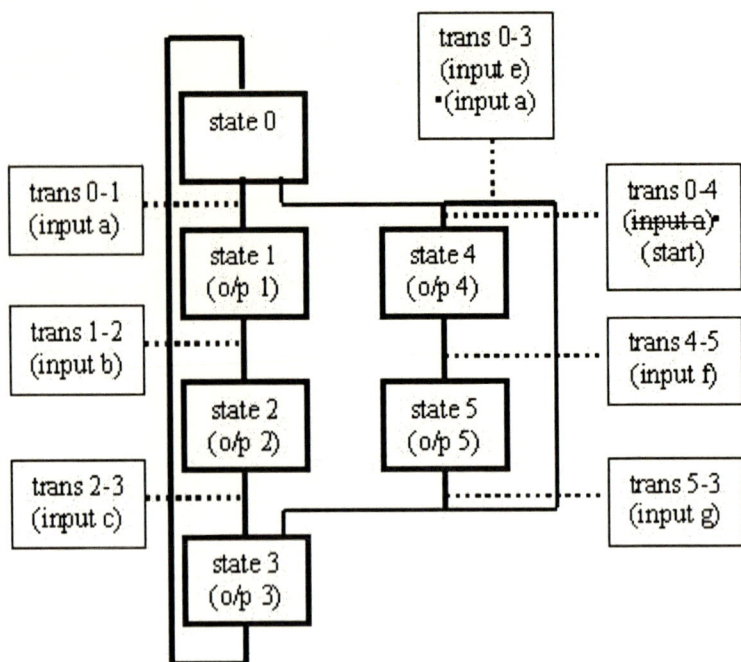

Fig 4.34

Select the only <u>correct</u> statement below regarding the diagram above.

Answer: ⌒   a. When state 0 is active, input a is off and the start input is on, state 4 becomes active

               ⌒   b. When state 0 is active and input e is on and input a is off, state 3 becomes active.

               ⌒   c. State 3 is entered directly from state 4

               ⌒   d. There are just two transitions leading to state 3

8.

The sequence for a set of traffic lights is to be plc controlled. The sequence is, red, red and amber, green, amber. Select the only incorrect statement below regarding the program.

Answer:
   ○   a. The transitions from one state to the next will be activated by timers.

   ○   b. Selective branching will be necessary for the red and amber part of the sequence.

   ○   c. The lights will be plc outputs.

   ○   d. Two outputs will be on during the red and amber state.

9.

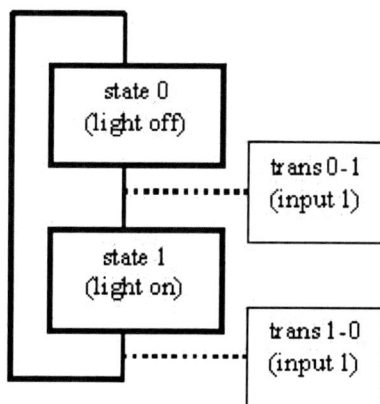

Fig 4.35

Select the only correct statement below regarding the diagram *Fig 4.35*.

Answer:
   ○   a. There can be no default state in such a sequence.

   ○   b. Input 1 only switches the light on.

   ○   c. Input 1 only switches the light off.

   ○   d. Input 1 alternately switches the light on and off.

10.

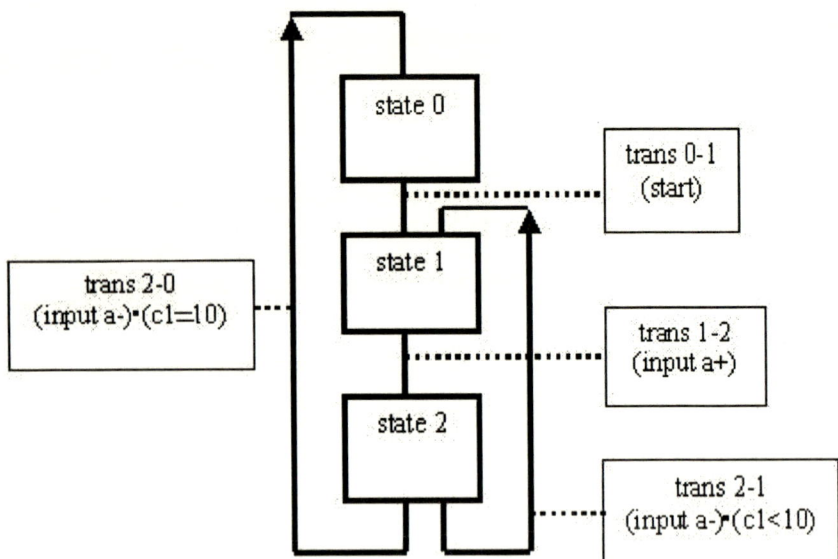

Fig 4.36

Select the only <u>incorrect</u> statement below regarding the diagram *Fig 4.36*. (c1 is a plc counter)

Answer:

    ○   a. If the plc count is less than 10, state 0 will always follow state 2

    ○   b. If the plc count is less than 10, state 1 will always follow state 2

    ○   c. Provision must be made in the system to reset the counter.

    ○   d. The counter present value is the deciding factor between trans 2-0 and trans 2-1

11.

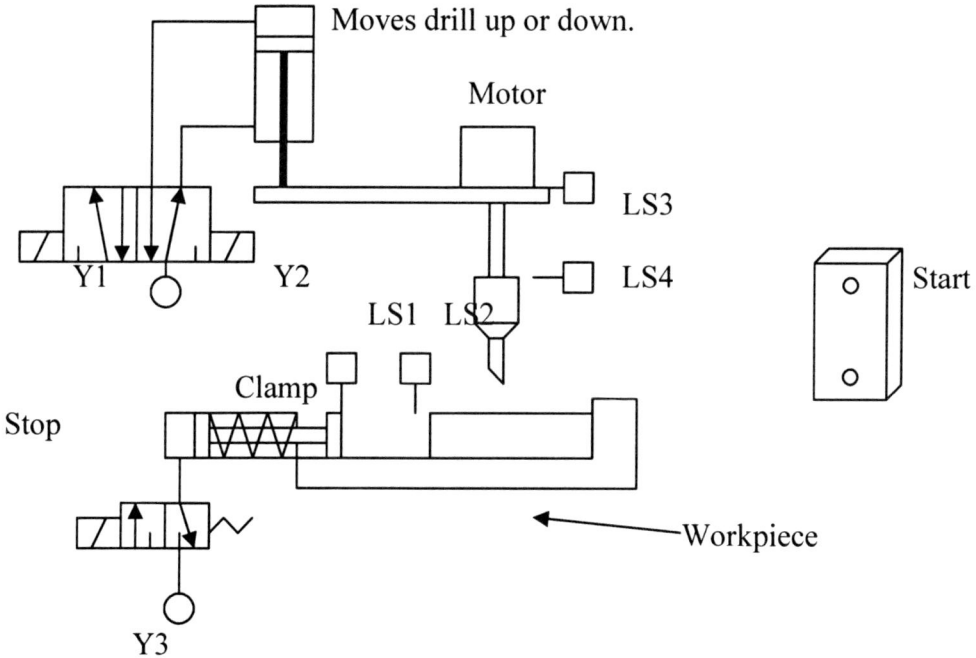

Fig 4.37

The diagram shows a workstation for the clamping and drilling of parts. When the start button is pressed the piece is clamped, drilled and then released.

If the stop button is pressed then both cylinders return to their retracted positions and the drill motor stops.

(a) Draw a PLC wiring diagram for the system.

(b) Draw a state/transition diagram for the sequence.

(c) Draw the necessary lines of ladder logic to operate the system as designed.

Chapter 5
Parallel Branching

5.1 Introduction

It is sometimes desirable in sequential systems to carry out certain operations simultaneously. This has the advantage of saving time during each cycle. The SF chart technique known as parallel branching allows for some states to be active simultaneously and so the events run in parallel.

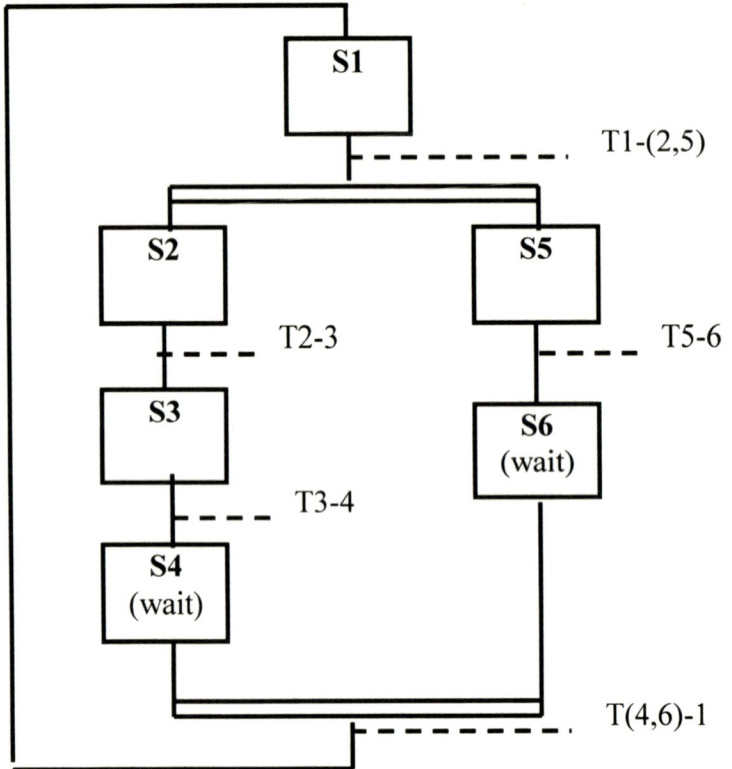

Fig 5.1

5.2 Parallel Branching and the SF chart:

Fig 5.1 shows a SF chart with parallel branching. The inputs and outputs have been temporarily omitted for clarity. The parallel section of the sequence is indicated by the double lines below S1 and below S4 and S6.

The transition T1-(2,5) energises sumultaneously states 2 and 5 after which the branches 2, 3, 4 and 5,6 proceed independently of eachother. The parallel section of the program only finishes when both branches are complete. To deal with this the final states in a parallel section (S4 and S6 in this case) are *waiting* states, similar to state 1, the initial state. When both waiting states are active T(4,6)-1 occurs and the sequence returns to S1.

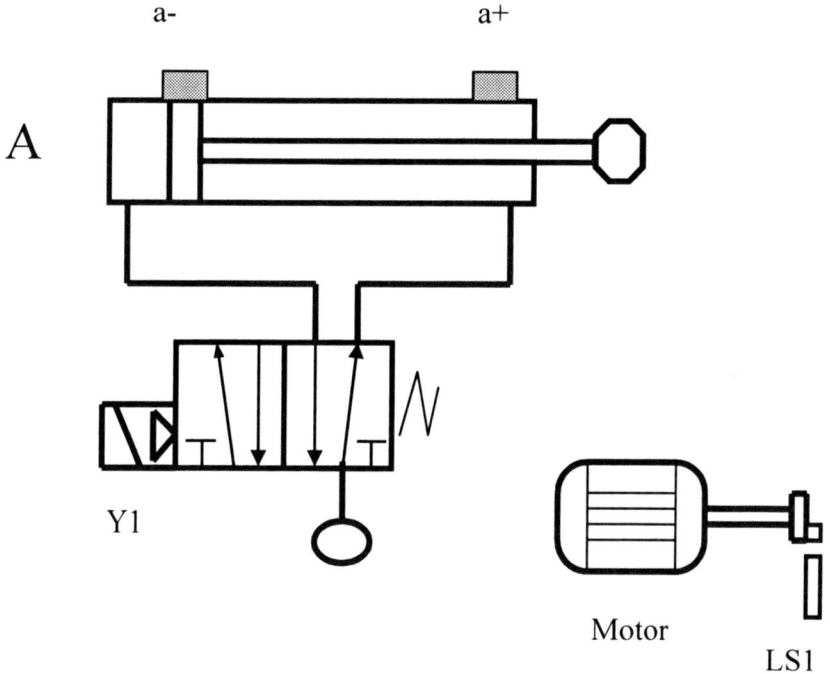

Fig 5.2

Example 5.1

When a *start* push button is pressed cylinder A in *Fig 5.2* extends and retracts once. At the same time the motor shaft rotates 5 times, each rotation being detected by the proximity switch LS1.

(a)   Draw a PLC wiring diagram for the system.

(b)   Draw a state/transition diagram for the sequence.

(c)   Write out an assignment list.

(d)   Draw the necessary lines of ladder logic to operate the system as designed.

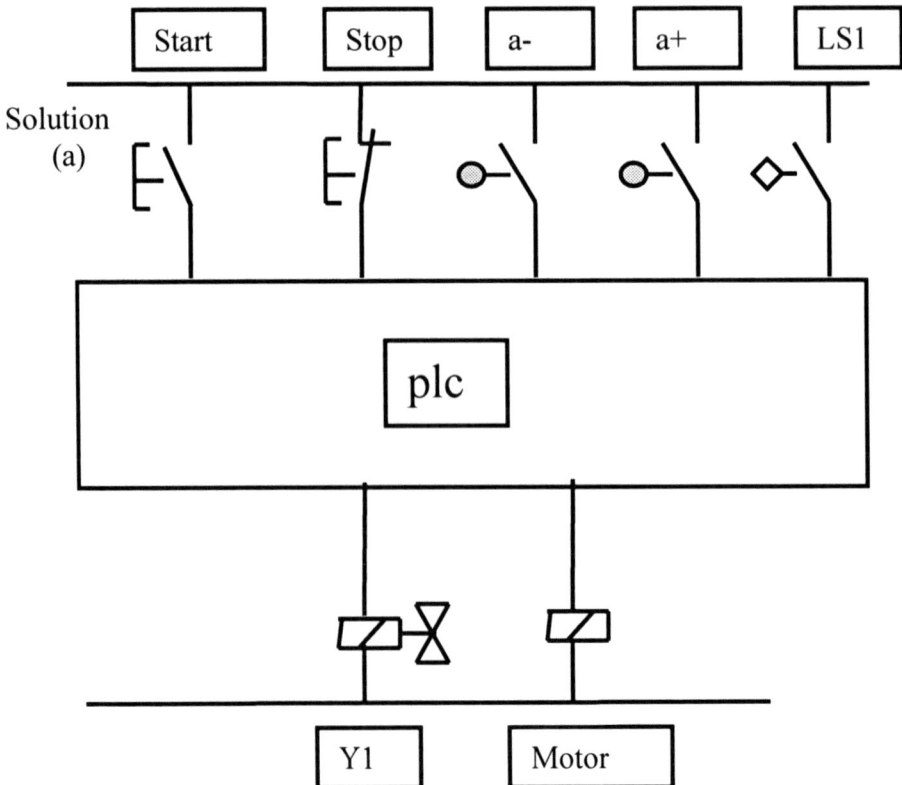

Fig 5.3

(b)

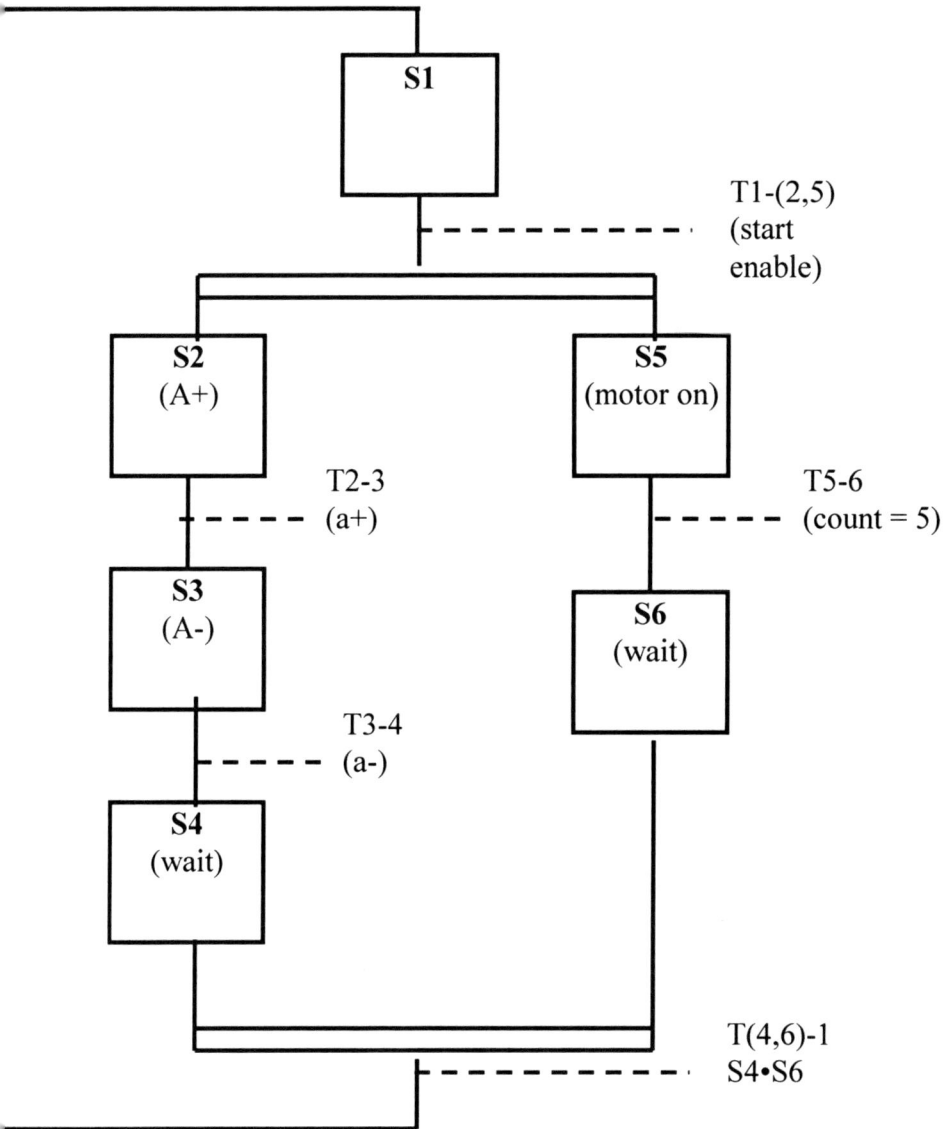

Fig 5.4

(c)

S1	R1
S2	R2
S3	R3
S4	R4
S5	R5
S6	R6
T1-(2,5)	R7
T2-3	R8
T3-4	R9
T5-6	R10
T(4,6)-1	R11
Start Enbl	R50

Table 5.1

(d)

Initial Conditions:

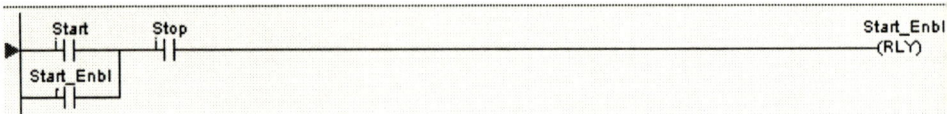

Fig 5.5

114

Transitions:

Note that the condition for T(4,6)-1 to be energised is that states 4 and 6 are simultaneously on.

Fig 5.6

States:

Two states (2 and 5) are switched on by T1-(2,5) and two states (4 and 6) are switched off by T(4,6)-1

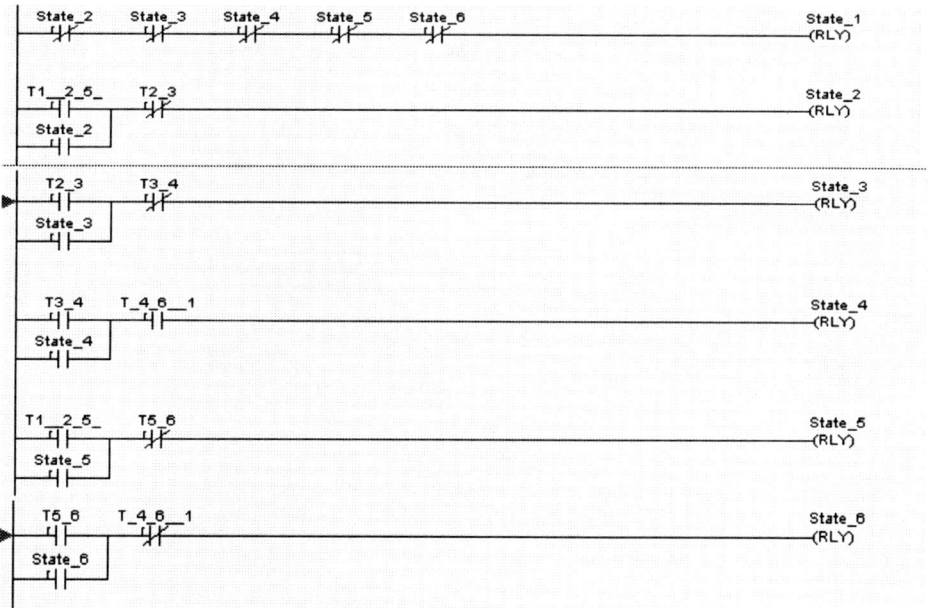

Fig 5.7

115

Outputs:

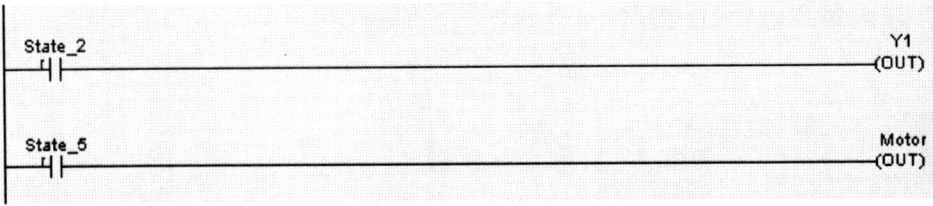

Fig 5.8

Counter:

While the motor is running, during state 5, the counter is incremented by the proximity switch LS1. The counter present value is reset to zero when the system is returned to state 1.

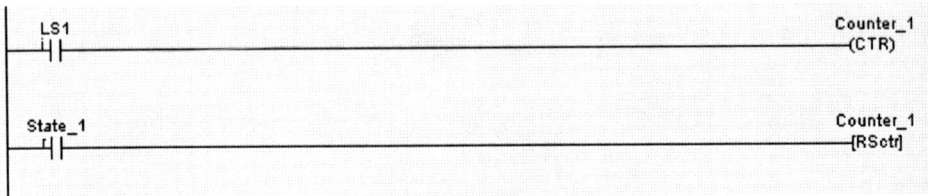

Fig 5.9

In the next example, we will design a circuit that uses parallel ranching and then, as a modification, we will add some selective branches to it.

116

Example 5.2

A drilling and countersinking operation is carried out on components on a conveyor using tools mounted on two pneumatic cylinders. The conveyor is indexed by a third cylinder extending and retracting. Each cylinder is operated by a 5/2 double solenoid directional control valve. The installation is shown in *Fig 5.10*.

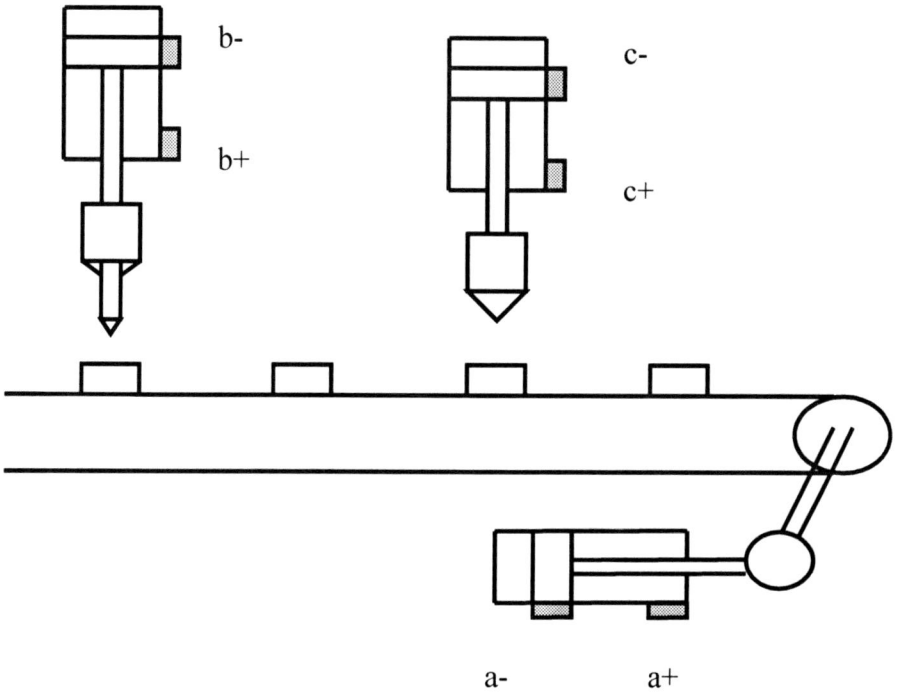

Fig 5.10

For the sake of simplicity no provision has been made for the clamping of the components. The task of inclusion of clamping has been added as an exercise at the end of the chapter.

The PLC wiring diagram and the SF chart are shown in *Fig 5.11* and *Fig 5.12*. The assignment list is shown in *Table 5.2* and the ladder logic follows.

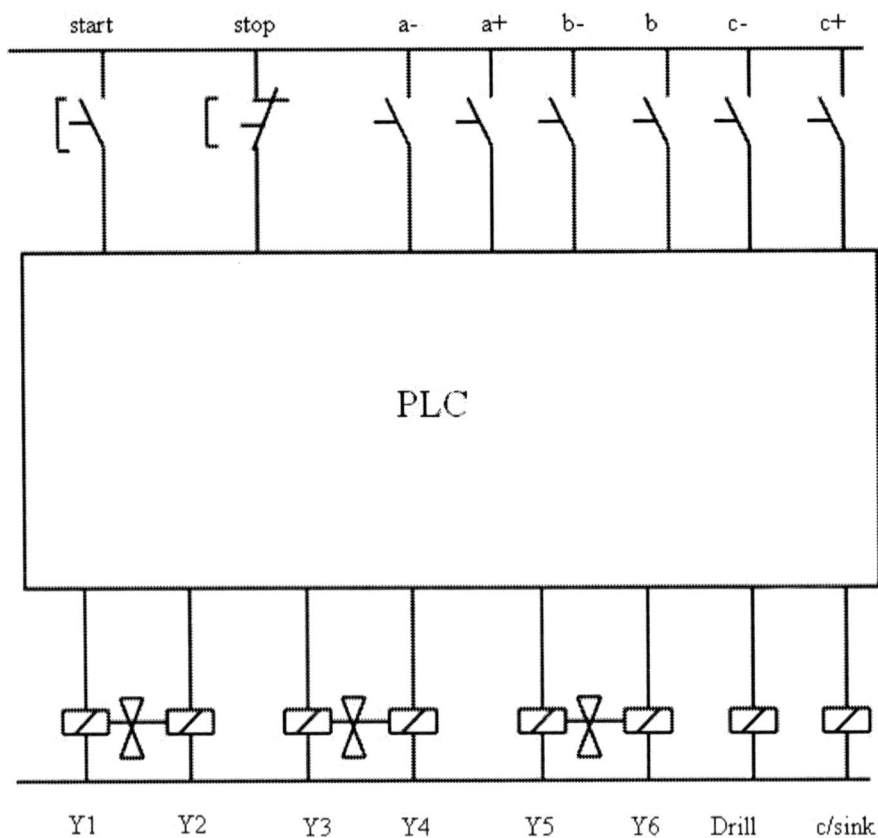

Fig 5.11

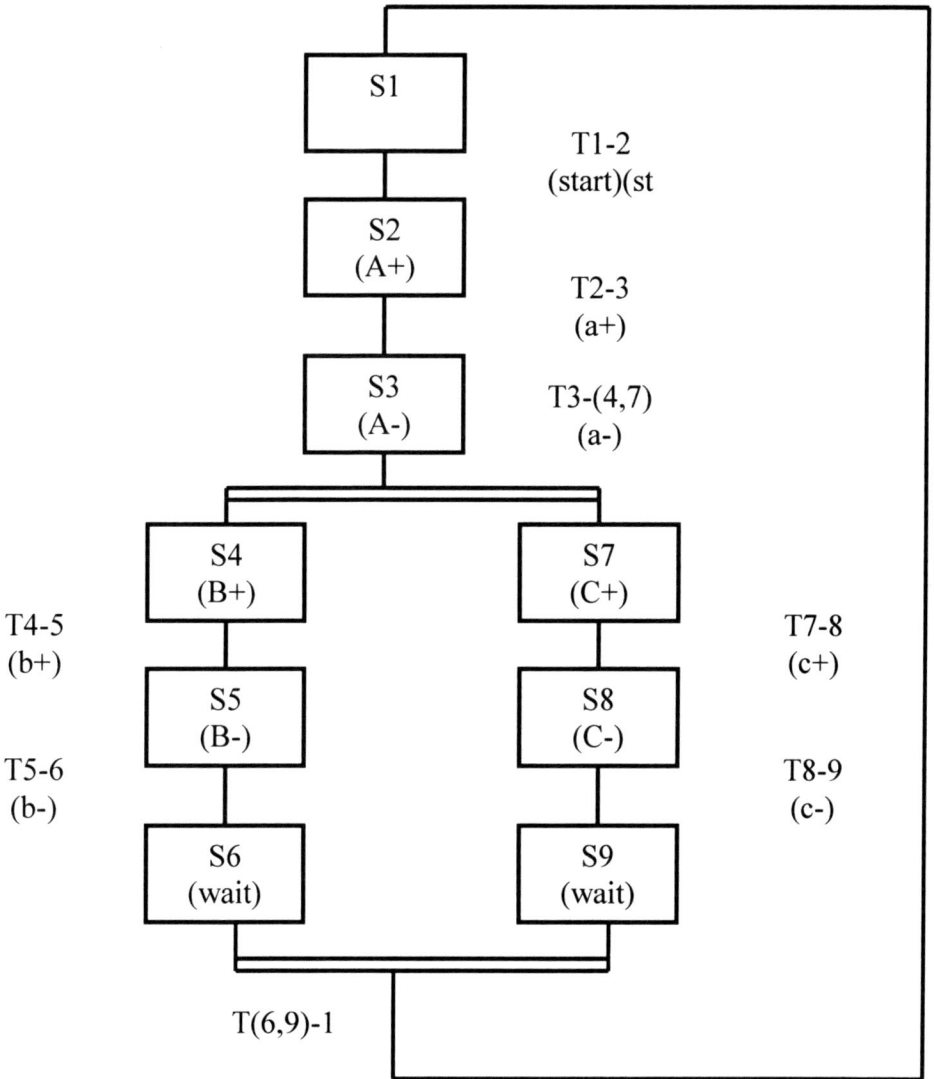

Fig 5.12

S1	R1
S2	R2
S3	R3
S4	R4
S5	R5
S6	R6
S7	R7
S8	R8
S9	R9
T1-2	R10
T2-3	R11
T3-(4,7)	R12
T4-5	R13
T5-6	R14
T7-8	R15
T8-9	R16
T(6,9)-1	R17
Start Enbl	R50

Table 5.2

Initial Conditions:

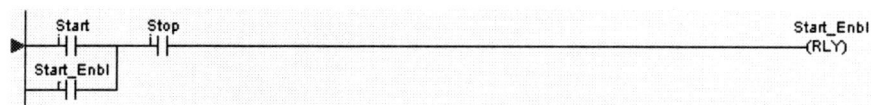

Fig 5.13

Transitions:

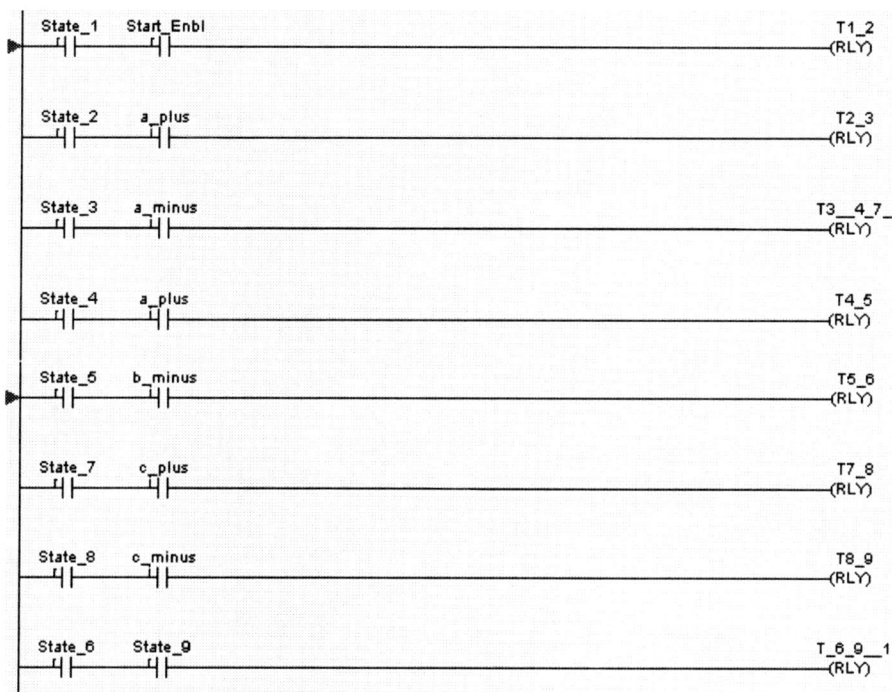

Fig 5.14

States:

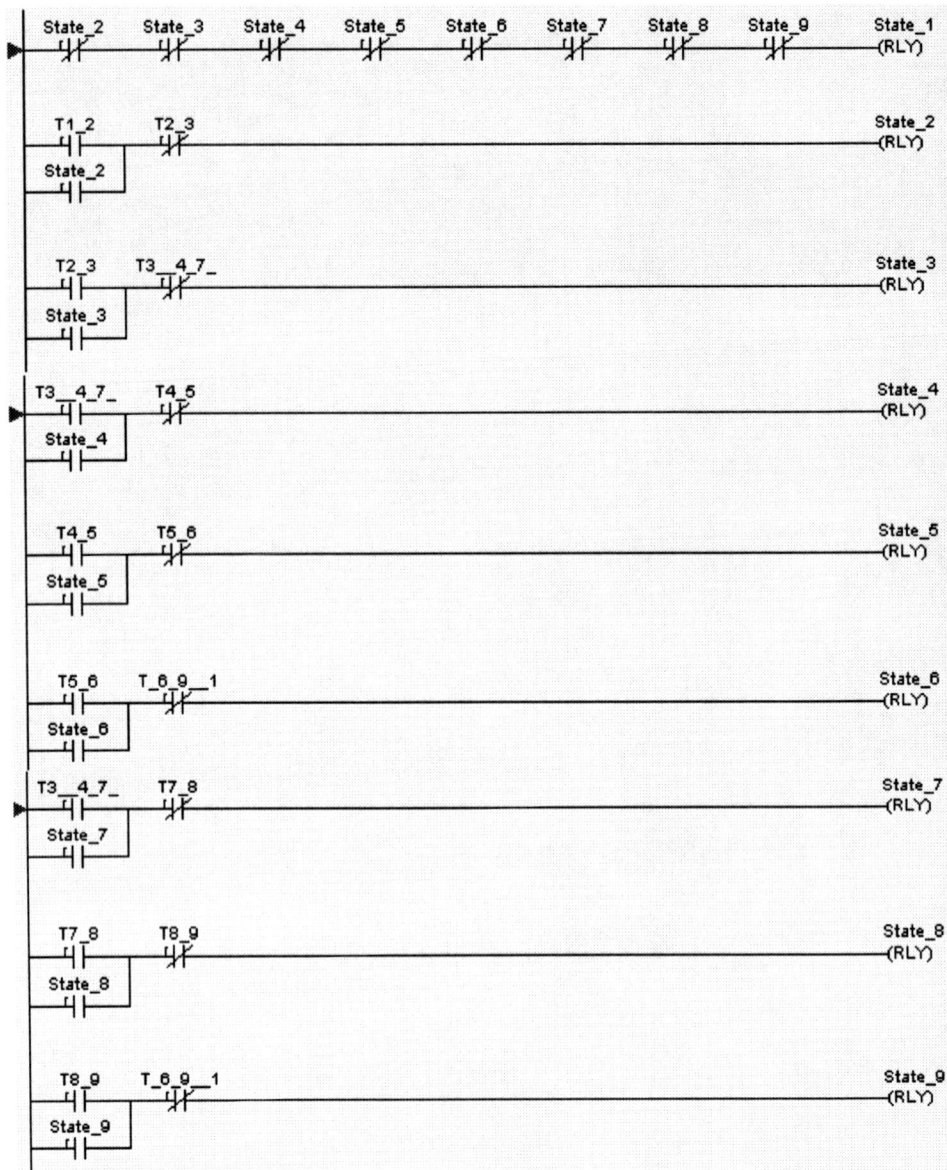

Fig 5.15

Outputs:

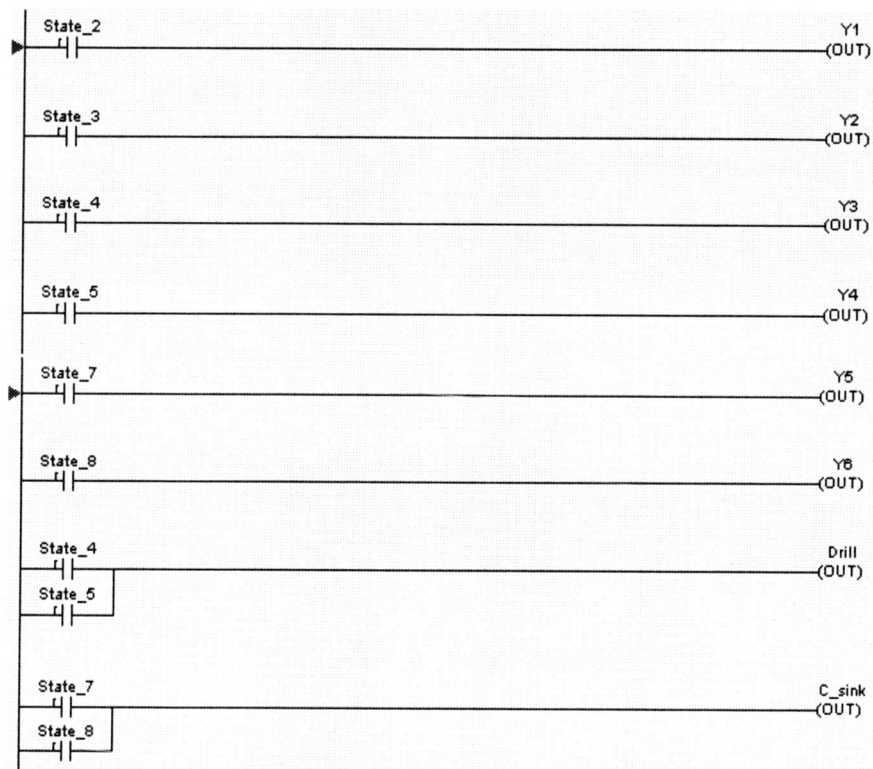

Fig 5.16

Example 5.3:

It has been decided to modify the drill and countersink process in Example 5.2 to allow for empty conveyor spaces. Two extra proximity switches are fitted to detect the presence of a part under the drill and under the countersink.

The modified SF chart is shown in *Fig 5.17*.

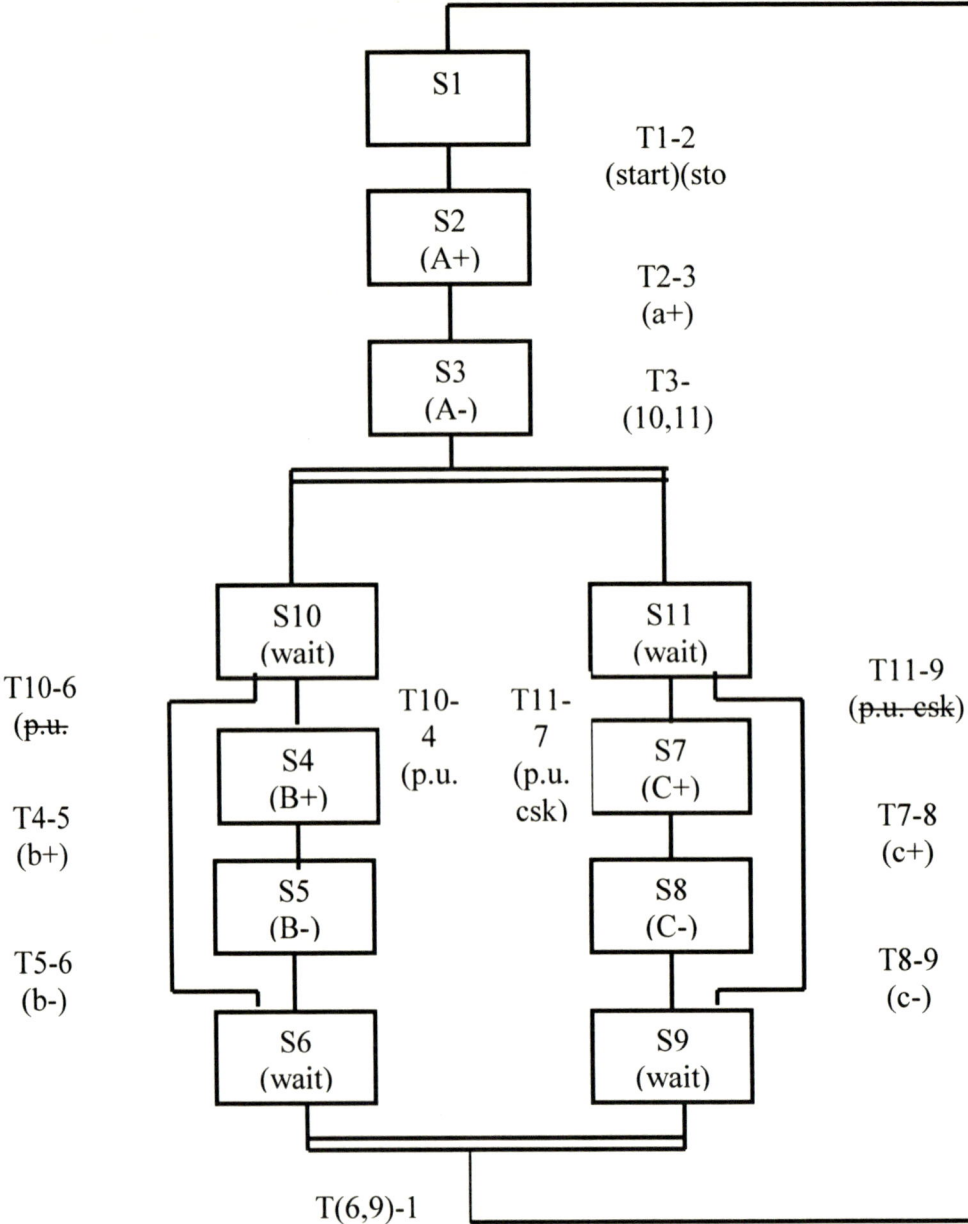

Fig 5.17

124

Two extra waiting states S10 and S11 have been introduced because it is not possible for a the sequence to simultaneously enter a selective and a parallel branch.

Once states 10 and 11 are active the choice of routes for the program depends on whether components are in place under the drill and countersink.

The PLC wiring diagram with the two additional inputs is shown in *Fig 5.18* and the modified assignment list is shown in *Table 5.3.*

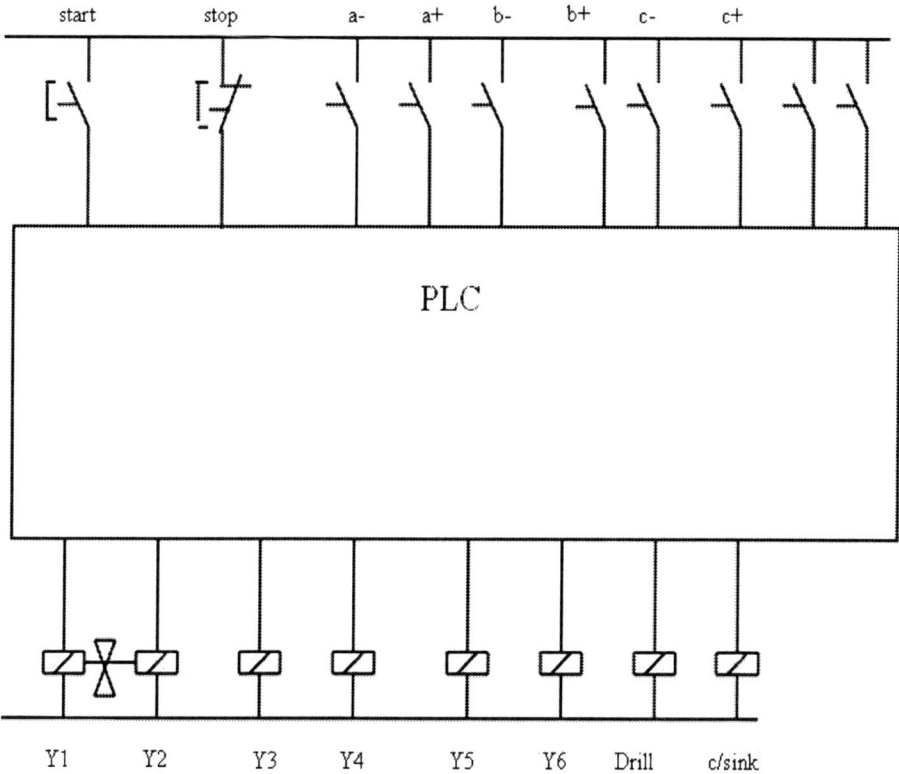

Fig 5.18

S1	R1
S2	R2
S3	R3
S4	R4
S5	R5
S6	R6
S7	R7
S8	R8
S9	R9
S10	R18
S11	R19
T1-2	R10
T2-3	R11
T3-(10,11)	R12
T4-5	R13
T5-6	R14
T7-8	R15
T8-9	R16
T(6,9)-1	R17
T10-4	R20
T10-6	R21
T11-7	R22
T11-9	R23
Start Enbl	R50

Table 5.3

Appendix
Using the Trilogi Software

The following gives instructions to getting started with the PLC simulator.

The software can be downloaded from http://www.tri-plc.com/trilogi.htm.

Open the plc Program Editor

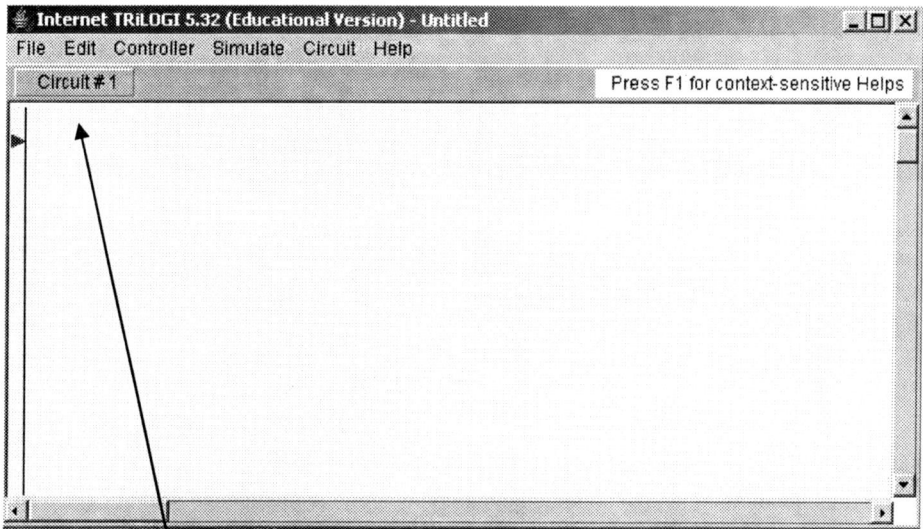

Click on the arrow to get the toolbar

Select normally open (n/o) input symbol

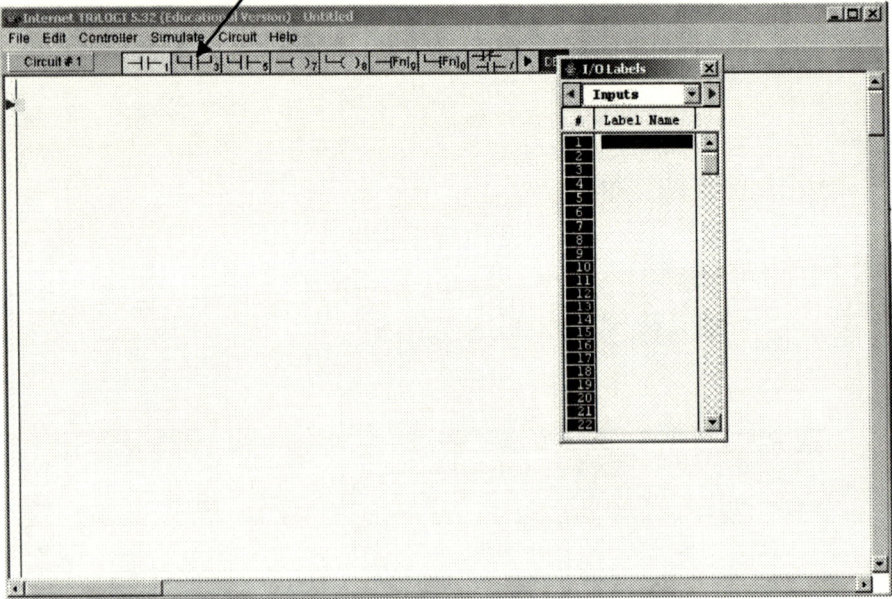

Click on the bar to get the
Define Label Name dialog box

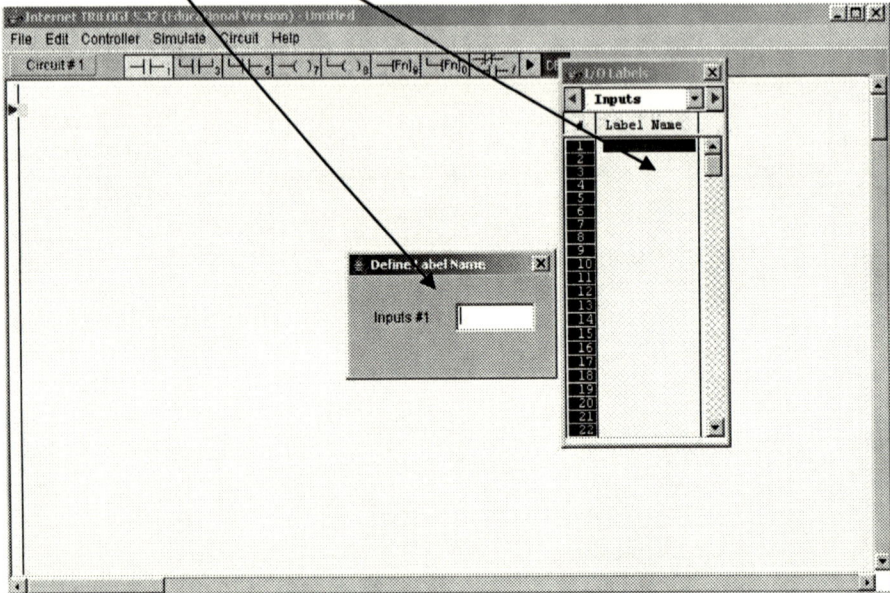

Call the input *Start_Button* or *Start Button* and press return.

Select another <u>n/o contact</u> and call it *Stop Button*

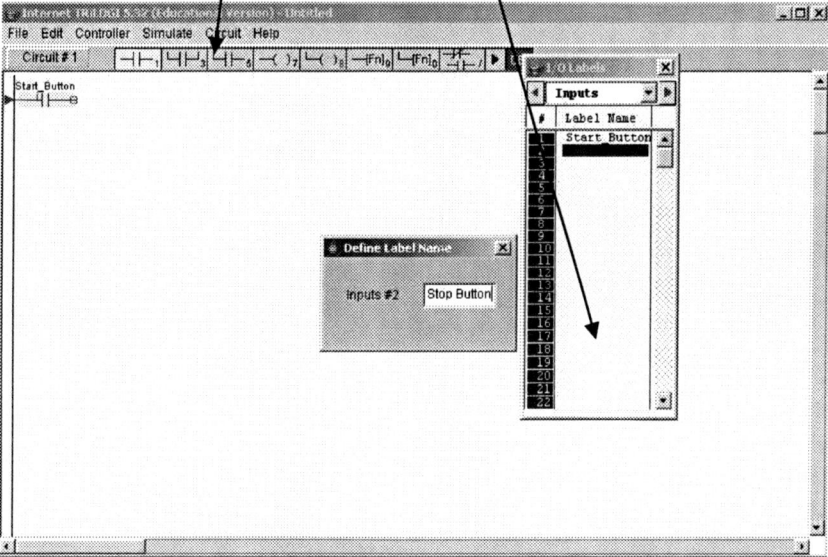

It appears in series with the *Start Button* contact

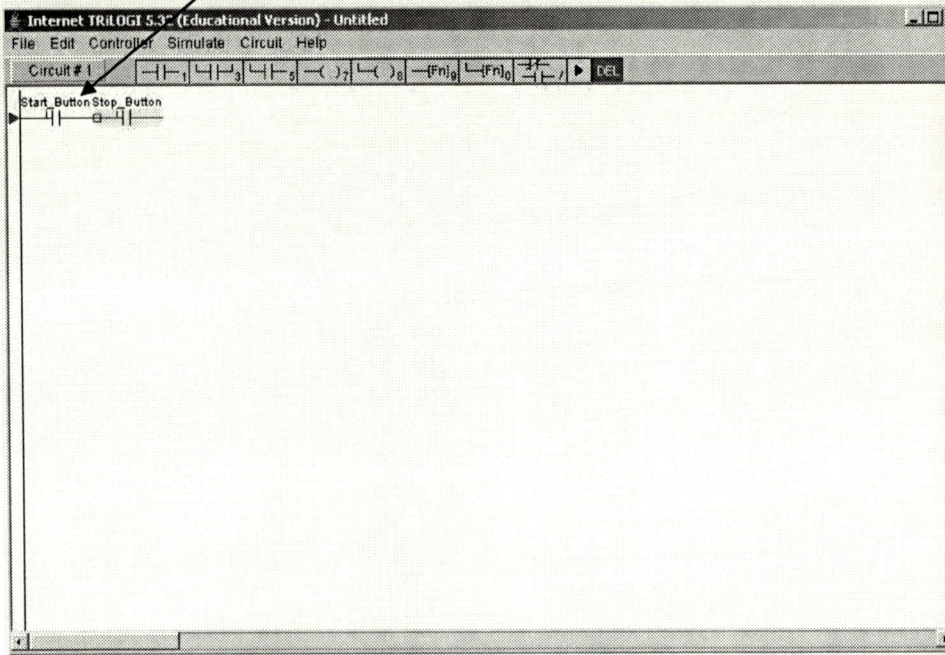

Clicking on the <u>right hand</u> box of the tool bar allows you to change the selected contact to n/c.

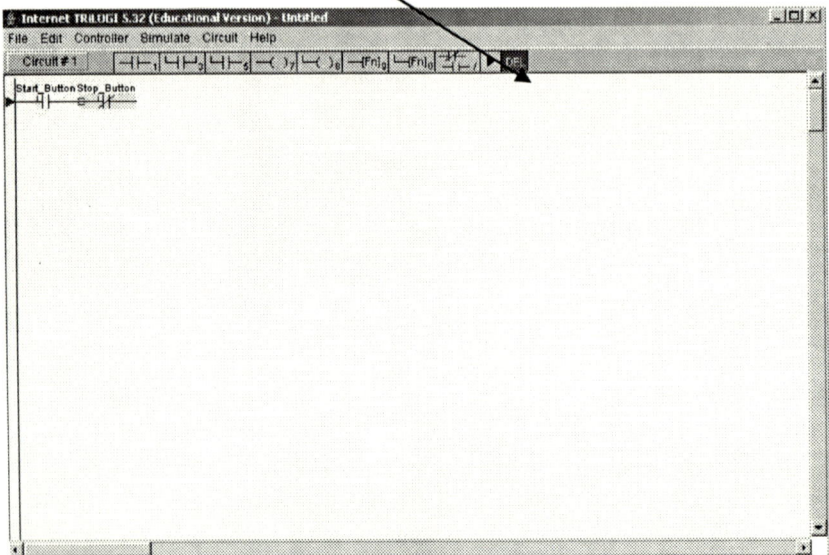

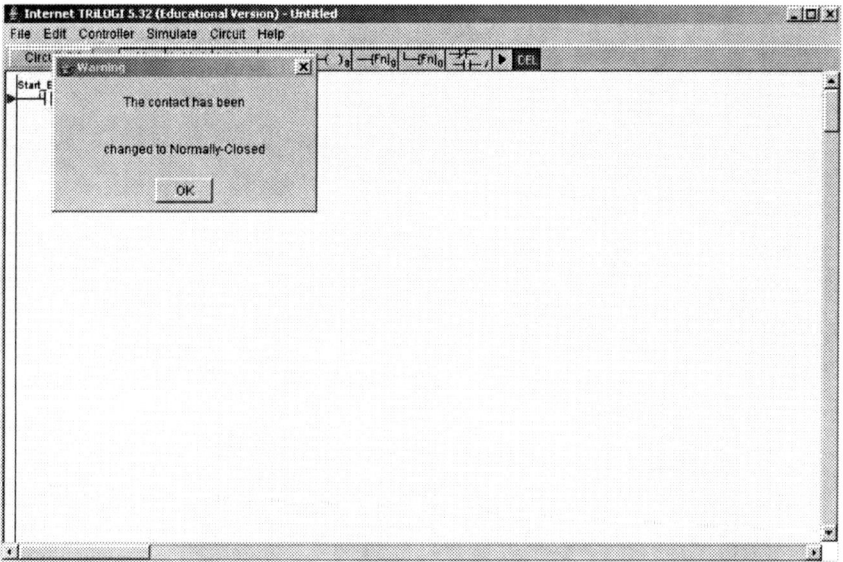

Now select an output and call it *light*

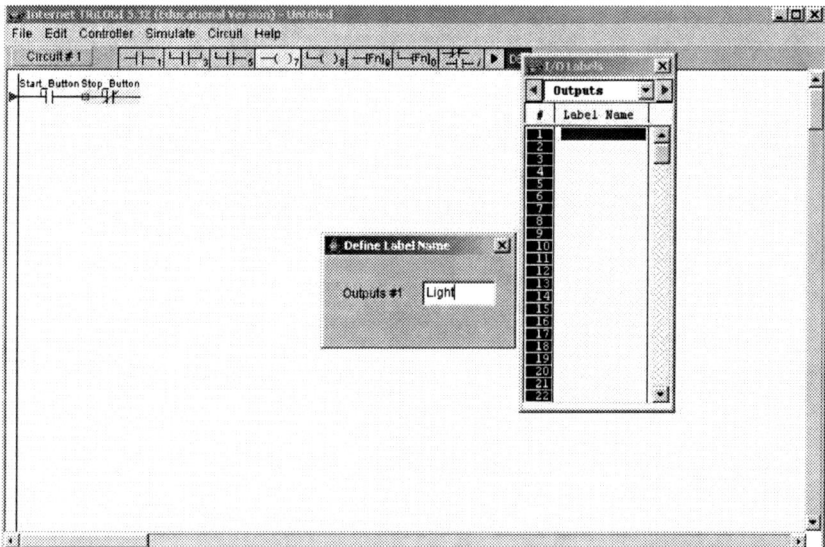

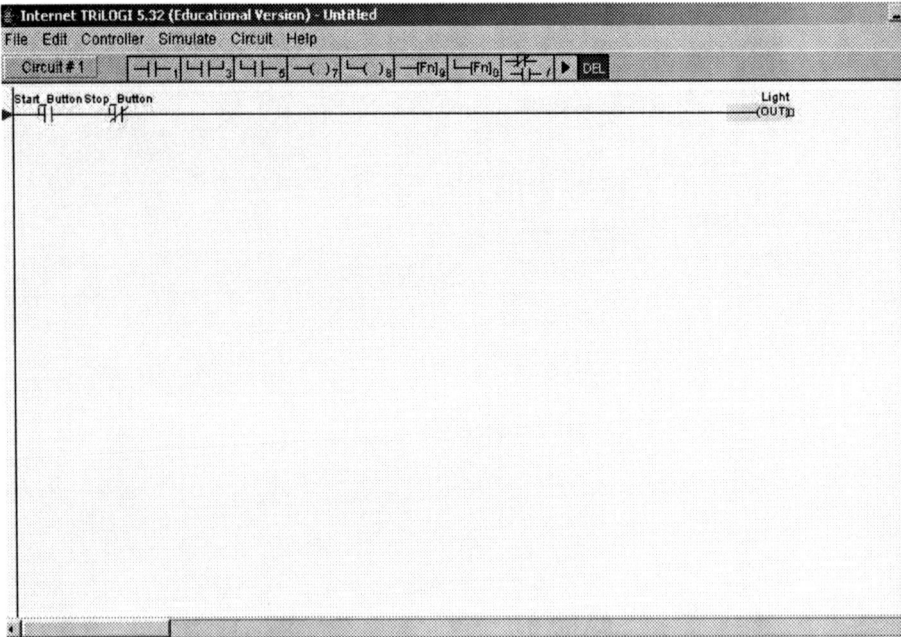

Highlight *Start_Button* again and select the <u>parallel contact</u> button.

Call the contact *light* and it will act as a latch for the light

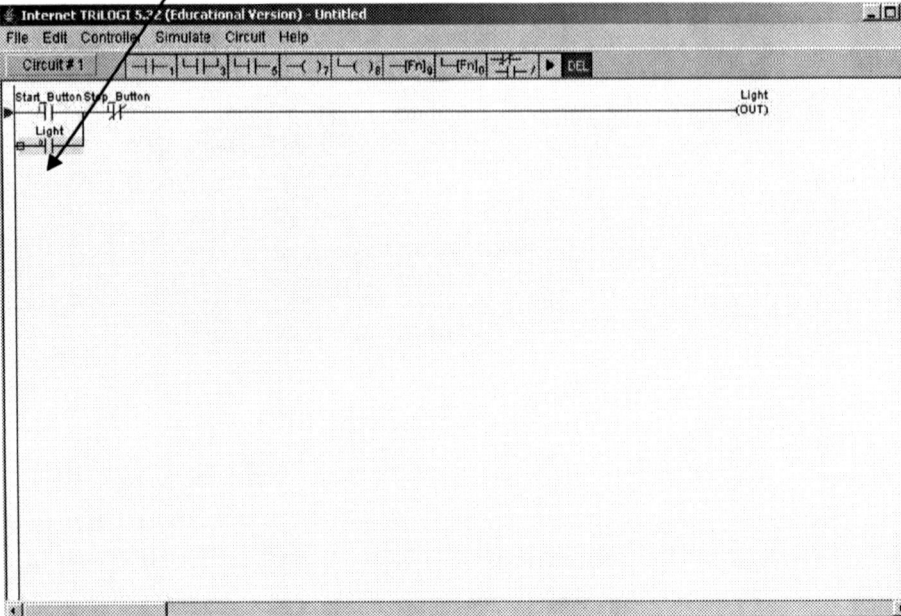

Click on the *Simulate* menu and select *Run (All I/O Reset)*

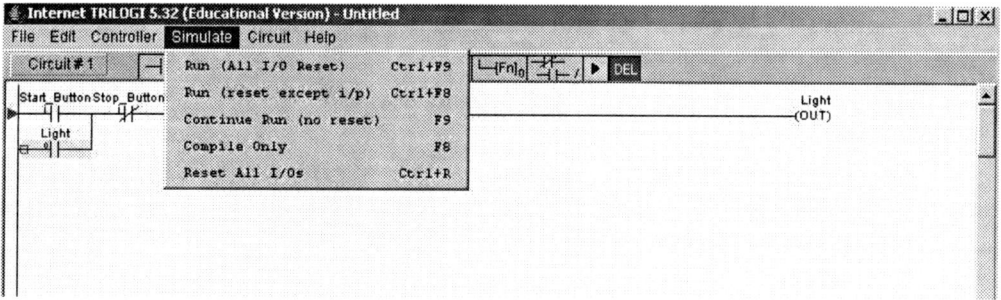

The two inputs and one output can be seen on the simulator.
Click on the *Start_Button* input to switch it on.

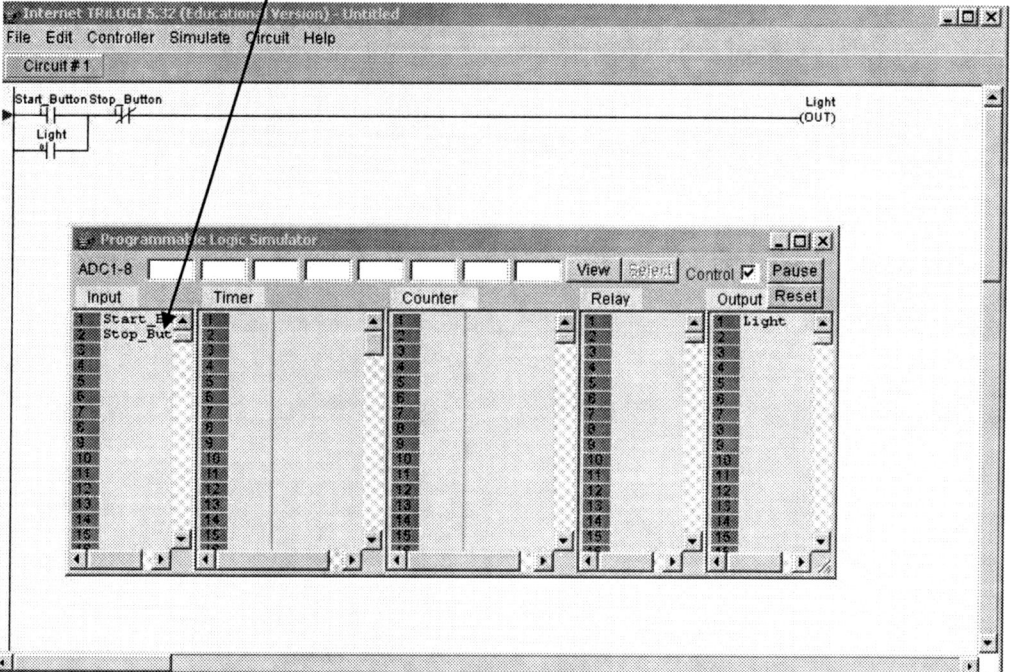

133

The output *Light* comes on and stays latched on, until the *Stop_Button* is pressed.

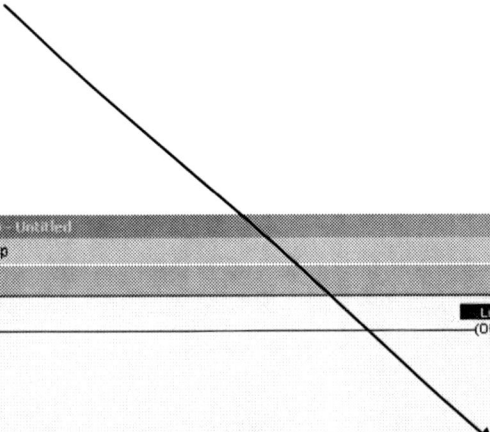

Printed in the United Kingdom by
Lightning Source UK Ltd., Milton Keynes
138255UK00002B/190/A